By the author of *Living Nonduality* and
One Essence, Robert Wolfe

SCIENCE OF THE SAGES

*Scientists encountering nonduality from
quantum physics to cosmology to consciousness*

Karina Library, 2012

ISBN-13 Print: 978-1-937902-04-9
ISBN-13 eBook: 978-1-937902-03-2

Karina Library
PO Box 35
Ojai, California 93024

I'm indebted for the assistance of Natalie Gray in manuscript preparation, and to Michael Lommel for design, editing and guidance. —RW

...a spirit is manifest in the laws
of the Universe—a spirit vastly
superior to that of man, and
one in the face of which we,
with our modest powers, must
feel humble.

—Albert Einstein

✧

Table of Contents

Introduction

In 1966, physicist Fritjof Capra received his doctorate at the University of Vienna. Meanwhile, he had "become very interested in Eastern mysticism, and had begun to see the parallels to modern physics." In 1976, Shambhala published his *The Tao of Physics*. Within a year and a half, it was in its fourth printing. It was then picked up by a book club, followed by a Bantam paperback which went into five printings in about two and a half years.

Thus, I came across it at a time when I too was becoming "very interested in Eastern mysticism," reading such spiritual teachers as Krishnamurti and Alan Watts (both of whom are named in Capra's flyleaf dedication).

The cover of the Bantam edition called it "A pioneering work." (A dozen "major publishers in London and New York" had declined the manuscript.) In some three and a half decades, an increasing number of quantum physicists and astrophysicists have come to recognize that "modern physics," as Capra puts it, "is harmonious with ancient Eastern wisdom."

Over the past twenty-two years, since my own awakening to the truth of the teachings of nonduality, I have steadily added to a list of such reading material. The scientists who are authoring these books are professional pragmatists, basing their judgments or conclusions on factual evidence. Whether one looks out at the mysteries of a vast cosmos or narrows the view to the counterintuitive behavior of a subatomic particle, I would not be alone in maintaining that nonduality is the basic principle that explains the Whole—the "spirit" of what is "manifest in the laws of the Universe," to borrow a phrase from Einstein.

Comprehending the nondual teaching, I know from experience, is not "rocket science." And understanding what physicists are reporting regarding such discoveries as "entanglement" is available even to magazine readers. What follows, here, brings together cutting-edge science revelations with revealing ancient insights clarifying ultimate Reality.

I am confident you will not fail to see the connection.

Robert Wolfe
May 1, 2012
Ojai, California

Prefatory Note

"It doesn't take an Einstein to understand modern physics." —Professor Richard Wolfson

Many people today have at least some acquaintance with the principles of physics, and even of some of the widely-reported aspects of quantum "mechanics," or theory— especially in the arena of experimental (as opposed to theoretical) physics. Astrophysics, by its nature, is less conducive to experiment, but the physics principles which are known to exist are applied in its study too.

Not every physicist or astronomer concurs with every fact gathered here, but there is broad general agreement on such facts within the parameters of what is known today.

Also, such factual material changes, from month to month, as new discoveries occur; but the basic principles of physics are not likely to change.

If you were a physicist or cosmologist, you would use the shorthand called exponential notation: rather than write out one trillion—1,000,000,000,000, involving twelve zeroes—you would write 10^{12} (or *say*, "ten to the twelfth power"). Similarly:

one thousand	10^3
one million	10^6
one billion	10^9

You could write the figure 1 as 10^0; this would allow you to write 1/10 (one tenth) as a negative figure of the whole number one (0.1): 10^{-1}. One thousandth then could be written 10^{-3}.

So, if you see 10^{15}, that's equivalent to 1,000,000,000,000,000. And 10^{-15} would read as .000000000000001; the size of a proton, for example, might be given in centimeters as 10^{-13}. A few real examples of exponential notation: number of cells in your body, 10^{13}; seconds elapsed since the Big Bang, 10^{18}; photons in the observable universe, 10^{88}.

Cosmic Birth

Cosmologists contend that the universe arose spontaneously out of absolutely nowhere as an utterly random act. Start "by imagining nothing, *don't* imagine outer space with nothing in it. Imagine no space at all."[*]

> "The Big Bang expansion is not an explosion in the classic sense, in which objects are flying out through pre-existing space like shrapnel. Space itself is expanding, stretching outward where it had not previously extended..."[†]

Imagine space arising from a singular point of unimaginable density maybe a *billionth* the size of a subatomic particle such as a proton. Imagine one second reduced to a negative *fraction* (imagine ten with thirty-four zeros following it), in which the initial point expands by twenty-five orders of magnitude—a pea growing to the size of our galaxy. "As the early universe went along doubling every microsecond, the stuff in it doubled, too—out of *nowhere*."[‡]

Quite literally, the universe began where you are now: not at some non-existent center of the cosmos far, far away.

---

[*] Brad Lemley, *Discover* magazine article (April 2002) concerning scientific conjecture on the origin of the universe.
[†] Kathy Sawyer, National Geographic (October 1999)
[‡] Lemley, 2002.

Even before the photons of light had materialized, the expansion was proceeding, according to science's calculations, at about a hundred times what would become the speed of light.

The end product:

> "...all matter and all the gravity in the observable universe indicate that the two values seem to precisely counterbalance. *All matter plus all gravity equals zero.* So the universe could come from nothing *because it is, fundamentally, nothing.*"[*]

The Big Bang "wasn't the emergence of the universe *into* space, but rather the emergence *of* space," according to physicist Brian Clegg in *Before the Big Bang*. Prior to the Big Bang, there was "not empty space; just *nothing.*"

All that we now know, concerning the size of the universe, is that we can see a *visible portion* of it, about thirteen and a half billion light-years in any direction—or some 27 billion light-years across, totally, "But that doesn't mean the universe *stops* at the limits of what it's possible to see." Even given the visible extent, our *planet* is an "infinitesimal speck"; even less so, as the universe doubles in size every ten billion years.

What is expanding is actually *space* itself. This creates an odd anomaly. At the farthest regions, "expansion of space makes it possible for light—*or* physical objects [such as galaxies]—to exceed" the speed of light in their movement

---

[*] Astronomy professor Mark Whittle elaborates: "The total mass/energy of the universe equals zero: the universe sums to nothing. This is comparable to what one associates with traditional spiritual-based cosmologies. This also gives us insight into how the universe came into being: perhaps it *came* from nothing."

with respect to each other, though that speed limit holds as a physical principle of limitation relative to space itself in our region of the cosmos, and as far as we know, the whole cosmos.

✧

"The Big Bang is simpler to understand than is almost anything we find here on Earth," states astronomy professor Mark Whittle.

"At the Big Bang, the expansion of space was infinitely fast." This is not an inert space; it is dynamic: "self-generating, self-sustaining." The term for it is "vacuum energy," now known also as "dark energy."

"Alive with quantum effects,"[*] in addition to qualifying as energy, it has weight; hence its mass creates gravity. Gravity pulls on mass. "Thus it's creating its own space."

Initially, the universe was "simply an ocean of a uniform, hot, glowing gas," an opaque fog.

A translation of the Taoist Chinese Prince Huai Nan Tzu, circa 150 B.C., says it well:

> "Of old, before the creation of Heaven and Earth, I consider there was the void without form or shape; profound, opaque, vast, immobile, impalpable and still: it was a nebulosity, infinite, unfathomable, abysmal, a vasty deep without clue of class or genera...."

Theoretical physicist Paul Davies:

> "...we can no longer think of a vacuum as 'empty'. Instead it is filled to capacity with *thousands* of

---

[*] Physics professor Robert March: "John Wheeler insists that during the early moments of the Big Bang, the universe was so small that quantum fluctuations must have played a major role."

15

different types of particles; forming, interacting and disappearing, in an incessant sea of activity....Nor is this quantum picture just an intellectual model. Very real physical effects occur, as a consequence of this fluctuating vacuum."

Due to fluctuating physical effects, expansion slowed over time; and with the emergence of atomic particles, the universe became transparent about four-hundred thousand years after the Big Bang.

The vacuum energy presently accounts for about three-fourths of everything in the universe. In 1998, it was determined to now be expanding at an accelerating rate again.

As Whittle puts it, "It is making more of itself." And as this space expands, it does not become more dilute, but (unlike matter) continues to maintain the same density. And because there is not anything outside of the space which is expanding (in other words, it is not ballooning out into a *pre-existing* space), there is not anything to stop its expansion. "There's something almost Zen about this."

"When you grasp the remarkable properties of vacuum energy," says Whittle, "you can't fail to be stunned."

He adds, "The universe is amazingly similar in every direction." When we look back in space to the origin of the Big Bang, we don't look in some particular direction. For as far as we can see, in *any* direction we look, we are looking at a remnant of the Big Bang.

This remnant is called microwave radiation, and it is essentially the signature of the left-over heat generated in the Big Bang. If you were able to see microwaves (as some of our telescopes and detectors can), you would be observing the diminished glow of the Big Bang wherever you looked out into the night sky.

If the universe were an eighty-year-old person, says Whittle, you would be able to see today (as our telescopes are able to do) back to the twelfth hour after that person's conception.

There are galaxies which can be seen with the naked eye that are three million light-years away. The Hubble Space Telescope, launched in 1990, has allowed us to image objects that are four billion times fainter than the faintest object that can be seen with the naked eye. Hubble has taken more than a half million images, such as those of galaxies that are 500 million light-years distant.

We have been able, in the last hundred years, to understand what was happening in the universe some thirteen billion years ago. But, due to the opaque primordial conditions, we will be blocked from seeing into the first 380,000 years after the Big Bang.

So, when we look back into space, for the origin of the Big Bang, the view is sealed off from us when we reach the point (in the space that expanded) where the foggy gas had not yet become transparent.

The space we are looking back over (in any particular direction) indicates a time span of 13.7 billion years to the origin of the Big Bang. In terms of the distance which light (the initial glow of the Big Bang) travels in a year, this suggests a distance in the universe of 13.7 billion light-years (as a form of measure).

However, space itself has been rapidly expanding during the past 13.7 billion years (when the radiating light was emitted).

Whittle: "...the universe extends well beyond our ('fourteen billion light-year') *visible* horizon. Current measurements indicate the universe's (curvature) radius is at least 150 billion light-years; but inflation theory suggests it may be *much* bigger." *

("Radius" would be a measure from our planet outward into any one direction. The same measure outward in the opposite direction would be "diameter"; in other words, double that of the radius.)

"The Earth shrinks into *insignificance* in the vastness of the universe."

As an average walker, you cover three miles in an hour. At this unrelieved pace, you would walk to the moon in nine years (238,857 miles away).

But if you traveled at the speed of light,[†] you'd reach the moon in a little more than 1¼ *seconds*.

The sun is 498 times more distant from us than the moon. It would take a flash of *light* 8⅓ *minutes* to be seen at the sun, from Earth.

Given a year's time, light traverses 5.8 *trillion* miles. The Hubble Space Telescope peers into (what we know of) the cosmos as far as 13 billion *light-years* distant.[‡]

---

* Writer Bill Bryson quoting Astronomer Royal Martin Rees: "This visible universe—the universe we know and can *talk* about—is a million million million million (that's 1,000,000,000,000,000,000,000,000) miles across. But according to most theories the universe at large—the meta-universe, as it is sometimes called—is vastly roomier still. According to Rees, the number of light-years to the edge of this larger, unseen universe would be written not 'with ten zeroes, not even with a hundred, but with millions.'"

† 186,000 mi./sec.; 11,160,000 mi./min.; 5.88 trillion mi./yr.

‡ We have observed stars whose light has taken 13.14 billion years to reach us—traveling, of course, at more than five trillion miles per year.

So, travel at the speed of light for a year (5.8 trillion miles) and travel for 13 *billion* years at that pace, and you will cover *only* the portion of the universe that we can "see" in one direction.

Princeton University cosmologist David Spergel: "We now know the age of the universe—13.7 billion years—to an accuracy of one percent."

And, he says, "four percent of the universe is atoms ["matter"] and ninety-six percent is something else, unidentified [so-called dark matter and dark energy]."

So, we do not even know what ninety-six percent of our universe is actually composed of!

Astrophysicist Charles Lineweaver and associate Tamara Davis, in a 2005 *Scientific American* article, outlined a number of modern discoveries concerning the cosmos.

"[In] modern cosmology, space is dynamic. It can expand, shrink, and curve, without being embedded in a higher-dimensional space.

"In this sense, the universe is *self*-contained. It needs neither a center to expand away from, nor empty space on the 'outside' (wherever that is) to expand into. When it expands, it does not claim previously unoccupied space from its surroundings....[The Big Bang] did not go off at a particular location and spread out from there into some imagined preexisting void. It occurred everywhere at once."

The cosmic expansion is still occurring—and accelerating. Galaxies, beyond a certain distance, recede from us (and other galaxies) faster than the speed of light; the only light that reaches us, from this horizon, is from some thirteen billion light years away. About a thousand objects have been observed passing beyond this (superluminal) horizon. Because it is space itself that is expanding, from the point of view of those objects, *we* are racing away from *them* at faster-than-light speed.

> "Imagine a light beam that is farther than the distance of 14 billion light-years, and trying to travel in our direction. It is moving toward us at the speed of light, with respect to its *local* space; but *its* local space is receding from us faster than the speed of light. Although the light beam is traveling toward us at the maximum speed possible, it cannot keep up with the stretching of space. It is a bit like a child trying to run the wrong way on a moving sidewalk. Photons, at the Hubble distance, are like the Red Queen and Alice, running as fast as they can just to stay in the same place.
>
> "....As a photon travels, the space it traverses expands. By the time it reaches us, the total distance to the originating galaxy is larger than a simple calculation based on the travel time might imply—about three times as large."

The universe is expanding at an accelerating rate. Theorists, says *Discover* magazine (February 2004),

"...suggest we're at the beginning of a very long process that will eventually result in what appears to be an *empty* universe. Trillions of years from now, matter will be so widely spread out that the average density will be much less than a single electron per quadrillion cubic light-years of space. That's so close to *zero* density that there's no meaningful difference."

Specifically,

"Stars will burn out, galaxies will disintegrate, and the universe will end, eternally dark and lifeless."

"A part of me is always surprised," says astronomer Patrick Petitjean: "I cannot stop asking, 'Why is the universe like this?!'"

The two words which probably most generally apply to cosmic objects are "unique" and "random." And the characteristic which most universally seems to apply to cosmic space is "emptiness."

If our *sun* was the size of a nine-inch *basketball*, its nearest planet Mercury—comparatively a little bigger than a poppy seed—would be orbiting 63 *feet* from the sun. Venus, then Earth (both smaller than a lentil) would be at 117 and 136 feet distant from the (basketball) sun. Mars (about half the size of Earth) would be some 750 feet (more than an *eighth* of a *mile* away) from the sun; and our outer-solar-system planet, Neptune, would orbit nearly a full mile from that nine-inch sun!

In terms of the content of what's between celestial objects, Corey Powell observes in *God in the Equation*:

> "If you scooped up a block of empty space 250,000 miles on a side—about the distance from the earth to the moon—you'd find just about one pound of energy inside, assuming you could find a magical technique for weighing it. In that same box of space, you'd find roughly half a pound of ordinary matter, mostly hydrogen atoms. The universe is very nearly empty."

It takes about eight minutes for the light from the sun to reach Earth; from the closest star, 4.3 years;[*] from the North Star, 460 years; from the nearest neighboring galaxy, 2.4 million years.

Our nearest neighboring galaxy, Andromeda, is nearly two and a half million light-years away. But we—Milky Way—and *it* are moving on a (long-term) collision course, at the speed of a quarter million miles per hour. Yet, given that most of space is empty, collisions among galactic objects will be rare.[†]

The distance from the sun to Earth (about ninety-three million miles), compared to the width of the *visible* universe,

[*] "Proxima Centauri, which is part of the three-star cluster known as Alpha Centauri, is 4.3 light-years away, a sissy skip in galactic terms, but that is still a hundred million times farther than a trip to the Moon." –Bill Bryson
[†] "The average distance between stars, out there, is 20 million million miles." –Bill Bryson

would be one millionth of a billionth of the universe's expanse.

Speaking of the emptiness of space, science writer Bob Berman, points to a bright star, Vega:

> "Vega is so distant that its light takes 25 years to reach Earth. But if you extended a one-inch-wide tube all the way from Earth to Vega and scooped up every bit of *matter* within, the contents would weigh just one-millionth of an ounce, roughly equal to a grain of sand."

So, to get an index to the extent of the visible universe:

> "...*scientists are confident that were you to weigh everything that's within our cosmic horizon, the tally would come in at about 10 billion billion billion billion billion billion grams.*"*

✧

In *The Infinite Book*, John Barrow cites an apt quotation regarding infinity:

> "Nor can one speak of [God] as having parts, for that which is 'One' is indivisible, and therefore also infinite—infinite not only in the sense of measureless extension, but in the sense of being *without* dimensions or boundaries, and therefore without shape or name."†

----

* Physicist Brian Greene.
† Early Christian text quoted in *The Infinite Book*, John Barrow.

As in the "not this, not that" tradition of the East, the definition of infinity basically tells you what it's not; without any limitations or boundaries. As such, the Infinite (or Eternal) has commonly been a stand-in name for the Absolute, or God. Being without limitation, it is said in the East that the Absolute can both be existent *and* nonexistent.

The notion of a god in *heaven* limits that figure to a particular locale: an *infinite* god would be equally present everywhere, being indivisible and having no definable parts: in heaven, hell, *and* on earth. Early biblical bookplates depicted God—*above* an orb representing (outer) the universe and (inner) the world—outside of both cosmos and earth, not present in, or permeating, these precincts.

With the use of the word "omnipresent" as a description of God in religious texts (and "infinite," "formless," "indivisible"), it became clear to some spiritual savants that there could be no definable central point anywhere at which one could deem God to be *explicitly* present (such as in a temple or church or mosque, or a heaven).

Holding just this sort of view (and proclaiming it), the Dominican-trained Giordano Bruno was burned at the stake, in Venice in 1600 A.D., by the Inquisition. Thus it seems somewhat ironic[*] that (less than four hundred years later), "In 1952, the Vatican embraced the picture of the expanding Big Bang universe as a natural conception of the Christian idea of creation out of nothing."

Barrow's scientific studies, of issues concerning the apparent infinity of the universe, lead him to state,

> "[Cosmic] expansion looks unstoppable. It will propel the universe into an ever-expanding future where all forms of life, no matter how complex or advanced,

---

[*] The Vatican now maintains an astronomy observatory in Arizona.

appear doomed to extinction....Acceleration to infinity sounds exciting, but it marks the end of everything that we value."

He quotes Olaf Stapleton, "Interference was included in His original plan."

Says Barrow, "Infinity...appears on the stage only when the crucial questions of existence are raised, [and] challenges us to contemplate...all that we hold dear."

Infinity constrains, by it's very nature, what we can know of it in our limited understanding. Though we might hypothetically learn whether the universe is actually infinite, such learning might take an infinite time. Barrows quotes Neils Bohr: "Prediction is very difficult, especially about the future!"

Hold a penny up to the sky, at arm's length, at any point you choose. Look through an adequate telescope, and (in far less than the radius of the penny) you will find *thousands* of *galaxies*, similar to our own (Milky Way). Look closely and you will discover that *each* galaxy contains *hundreds* of *billions* of stars (like our sun). Look around the entire sky and you will observe about *a hundred billion galaxies*. Some authors say, double that number.*

A hundred years ago, we didn't even know that there were other galaxies! Now we know, for instance, that our own Milky Way galaxy is composed of about a hundred billion

---

* Astronomer James Geach: "Scaled up to the whole sky, such a density implies a total of 200 billion or so galaxies. And those are just the most luminous ones; the true number is probably much larger."

stars (like our sun). Cosmologists remind us that this quantity is about as many grains of sand as would fill a cubic meter (close to forty inches per side). Our neighboring galaxy, Andromeda, has about three-hundred billion stars; and there is a galaxy estimated to have eight-hundred billion.

On a comparative scale, if Earth's orbit around the sun was reduced to the size of a pinhead, our galaxy would be about as wide as the United States.

If our galaxy (a hundred thousand light-years across) was reduced to the size of the United States, its stars would be the size of human cells—each separated by the length of a football field.

Reduce our galaxy to twenty yards across, and Andromeda galaxy would be another six hundred yards distant. (And, at that *scale*, the distance to the limit of the visible universe, in any direction, would be approximately 2,500 miles.)

The closest big galaxy to our own is Andromeda (just visible to the naked eye). How close? A spaceship traveling from here, moving at light speed, could reach only halfway there in a million years.

You're moving, standing on Earth, at about 795 miles per hour, breaking the sound barrier. And that's just the eastward spin of the planet on its axis. Meanwhile, Earth is in orbit around the sun at about sixty-seven thousand miles per hour. And, then, the solar system is rotating around the core of our galaxy, about every 250 million years, at some 518 thousand miles per hour—about one thousandth of the speed of light. Too, our galaxy is in motion toward Andromeda galaxy at around 288 thousand miles an hour. More yet, our local *group* of galaxies is speeding toward the constellation Virgo. "Alien astronomers in a galaxy a hundred million light-years away," says science writer Bob Berman, "would see us whizzing in the opposite direction" at more than five million miles per hour.

"Or are *they* moving away from us? The usual interpretation is that the space between us is increasing, so everybody is moving and yet nobody is actually moving. That's another way of saying that there is no center to the Big Bang. It happened everywhere and nowhere. Perhaps all we can say for sure is that we've come a long way, yet we're still going nowhere—fast."

About *325* million light-years from our galaxy, the Coma *cluster* contains many thousand individual galaxies, orbiting one another, with most of the galaxies containing more than a hundred billion stars (suns, somewhat similar to ours). The cluster is several million light-years across.

According to theoretical physicist Gabriele Veneziano: "As you play cosmic history backward in time, the galaxies all come together to a single infinitesimal point, known as a singularity—almost as if they were descending into a black hole. Each galaxy (or its precursor) is squeezed down to zero size. Quantities such as density, temperature and spacetime-curvature become infinite."

Four galaxies, in a particular cluster of galaxies, are in a collision which will result in a merger that is ten times as large as our Milky Way galaxy.

One observed star cluster contains several stars of surprising size, including one weighing 265 times as much as our sun. At 165,000 light-years away, it shines as bright as ten million suns of our type.

There is a galaxy, which is some 110 million light-years away, that has arms in a counterclockwise configuration, yet it is rotating in a clockwise direction.

"Once a second, somewhere in the universe a star explodes," says an article by Ron Cowen in *National Geographic* (March 2007), "blazing as bright as hundreds of billions of stars," or an entire galaxy.*

And: "A *nearby* supernova—within a *few* light years—would bathe the Earth in lethal radiation"; the implosion of a sun is actually a nuclear explosion, of a high order. The heat generated is "a hundred *billion* degrees"; and, "For someone brave enough to come within hearing distance, the waves would be audible, roughly the F note above middle C."

For a supernova in or near our galaxy, we'd be subject "to a big, big noise." A supernova occurs in the Milky Way, on average, every hundred years.

----

* A supernova can be as much as 160 thousand light-years across, and visible with the naked eye.

A report in the Los Angeles Times (7-26-03) gave the calculation of an Australian astronomer and his team, stating: "There are approximately seventy sextillion—that's 7 followed by 22 zeros—stars in the *known* universe."

A star "cluster"—not even a galaxy—can contain about a hundred thousand stars. (Daylight would be constant in such a location.)

A leaf of grass is no less the journeywork of the stars.
— Walt Whitman

"Human beings are made of...stardust," says Joel Primack in a book co-authored by his wife, Nancy Abrams: *The View from the Center of the Universe.* The iron atoms in your blood, carrying oxygen to your cells; the oxygen itself; most of the carbon in the carbon dioxide you exhale, for starters, owe their origin to exploding white-dwarf stars, detonating supernovas, planetary nebulas, and other violent phenomena.* Some of the explosions of massive stars (as supernovas) occurred even before our solar system was formed (close to five billion years ago).

Since that time, we have five thousand years of recorded human history, which represents only one millionth of the history of the Earth. During the time that the Bible was

---

* Our sun, at present, is creating carbon and oxygen.

29

written, there was not even the mathematical *means* to comprehend such things as the age of the universe.

Right now, we can see more galaxies than we'll ever again be able to see, because many are disappearing (due to cosmic expansion) beyond the outer horizon which is within our view.*

The particles that form your body have been around for billions of years. They are products of a universe which, scientists have determined, is composed seventy percent by something called dark *energy*, which we as yet have learned little about. Another twenty-five percent of the universe is comprised of dark *matter*, which likewise we know little of. Then another four percent is matter with the composition of atoms, but which is not illuminated in a form visible to us. All the atomic matter which *is* visible to us (galaxies, stars, planets, comets, debris) represents only about one half of one percent of the balance. So, all of matter as we know it, makes up only as much as five percent of the cosmos!

And so, closer to home, we have matter and antimatter among the products of the Big Bang, after the cooling of the crucible that was so intensely hot that "it makes no difference if it's Fahrenheit, Celsius or Kelvin." Nor did the speed of light make any difference at a time when developing processes preceded the existence of photons, thus light itself.

And yet, what can we say definitely "exists" today? Even the Big Dipper which man has known throughout history is in flux, as the stars which compose it move in relation to each other. Similarly, what astronomers call "the Virgo cluster of galaxies" is a mental *construct*; the galaxies that we see as a "cluster"—the light from each arriving in real time—radiated their light from different times in cosmic

---

* A hundred billion years from now (give or take a year or two) there will be no galaxies visible beyond our own.

history. Even the Earth's rotation has not been *constant*, having rotated much faster in its earlier stages.

Primack and Abrams comment, "Einstein (and many other scientists) have shown us that things are not as they *seem*."

A galaxy can have a mass that is equivalent to more than a hundred billion "solar masses"; a galaxy cluster, hundreds of trillions of solar masses. A black hole, in a galaxy cluster, may be only hundreds of millions of solar masses; but it may have (says a *Scientific American* article) "gulped down the equivalent of three hundred million suns, in the past hundred million years."

The Hubble Space Telescope surveyed forty galaxies, around the year 2000, and all had a black hole at their center. A black hole can be as large as ten million times the mass of our sun, and swallow two billion sun-size stars.

Time magazine (Frederic Golden: 6-25-2001):

"Now about halfway through its estimated 10 billion-year lifetime, our sun is slowly brightening. In about one billion years, its energy output will have increased at least 10%, turning Earth into a Venusian hothouse where plants wither, carbon dioxide levels plummet, and the oceans boil off."

And this is just a long-term prelude to things to come, writes Michael Lemonick, in the same science article.

"By the time the final chapter of cosmic history is written—further in the future than our minds can grasp—humanity, and perhaps even biology,

31

will long since have vanished. ...Finally, though, all that will be left in the cosmos will be black holes, the burnt-out cinders of stars, and the dead husks of planets: the universe will be cold and black. ...Eventually, even these will decay, leaving a featureless, infinitely large *void*. And that will be that. ...

"What we call the universe, in short, came from almost nowhere in next to no time," and to emptiness it is predicted to return. "The universe, once ablaze with the light of uncountable stars, will become an unimaginably vast, cold, dark and profoundly lonely place."

Though a black hole does not contradict Einstein's cosmic predictions, he did not believe such a thing could exist. A Cambridge University astronomer (Andrew Fabian) studied for a decade (according to a July 2002 *Discover* magazine article) a galaxy 130 million light-years distant, where a black hole has been calculated to be as big around as the orbit of Mars around our sun.[*] Information about it has been received by NASA's five-ton Chandra X-ray Observatory which has an elongated orbit that swings it (about six thousand miles above Earth) to nearly eighty-seven thousand miles into space. So, due to telescopic reports, "it would be hard to find a physicist or an astronomer who doesn't believe in black holes," says article writer Robert Kunzig.

A black hole is "an infinitely deep hole in the fabric of four-dimensional space-time; it forms when a massive star implodes" until all its mass is concentrated in a singularity—a point far, far smaller than a subatomic particle. "At this

---

[*] *Science* magazine (1-21-11) reports a black hole, with a mass of 6.6 billion suns, which is four times as large as the orbit of Neptune.

point, space-time ends, and the pull of gravity becomes infinite." According to astrophysicist Mitchell Begelman, "Space isn't sitting there stationary outside the hole." Even space-time, and the light that's in it, is being swallowed by the hole—in addition to matter that's in its vicinity. Astrophysicist Andrew Hamilton on black hole details (via Steve Nadis in *Discover*, June 2011):

> "Black holes are massive objects that have collapsed in on themselves, creating a gravitational suction so intense that their insides become cut off from the rest of the universe. A black hole's outer boundary, known as the event horizon, is a point of no return. Once trapped inside, nothing—not even light— can escape. At the center is a core, known as a singularity, that is infinitely small and dense, an affront to all known laws of physics....

> "A black hole, Hamilton realized, could be thought of as a kind of Big Bang in reverse. Instead of exploding outward from an infinitesimally small point, spewing matter and energy and space to create the cosmos, a black hole pulls everything inward toward a single, dense point."

And physicist Brian Greene:

> "It is common to speak of the center of a black hole as if it were a position in space. But it's not. It is a moment in time. When crossing the event horizon of a black hole, time and space (the radial direction) interchange roles. If you fall into a black hole, for example, your radial motion represents progress through time. You are thus pulled toward the black hole's center in the same way you are pulled to the

next moment in time. The center of the black hole is, in this sense, akin to a last moment in time."

The black hole of Andrew Fabian's study is said to now have a mass one hundred million times that of our sun, with a circumference of more than a hundred million miles.* This compares to the black hole that appears to exist at the center of the Milky Way galaxy which is estimated to weigh in at only 2.6 million suns. "There may be millions of black holes floating around our own galaxy, each five or ten times as massive as our sun, and roughly fifty miles around." The "event horizon" of a black hole may be, according to estimates, six miles across or even six-thousand light-years across.

As for our galaxy's black hole in relation to our solar system, we are twenty-seven thousand light-years away; but there are a hundred thousand other stars which are within a light-year of it.† Consequently, according to Ken Croswell (*National Geographic*, December 2010), "Every now and then, the black hole swallows...a wayward planet,‡ or even an entire star." This happened as recently as the mid 1600s, and again in the 1940s, research says.

"Surprisingly, the black hole also catapults stars away," one observed streaming away into intergalactic space at more than a million-and-a-half miles per hour. "The black hole may have flung a million stars out of the galaxy, in this fashion."

---

* Evidently it developed during our Cretaceous Period, more than a million years ago.
† One light-year is 5.88 trillion miles.
‡ It is calculated that a black hole can swallow one Earth-size mass every two minutes.

Not until 1990 did we know of other planets around other stars than our own. The indications are that more than half of the stars in our galaxy have planets around them.*

NASA has announced (L.A. Times, 2-3-11) the result of its study of nearly a thousand stars, in a band between five-hundred and three-thousand light-years distant: 1,235 planets were detected, but only fifty-four of those would be positioned (in relation to their sun) to possibly have liquid water; only five of these approach Earth's size.

California astronomer and exoplanet specialist Geoff Marcy estimates that (writes Michael Lemonick) "our *galaxy* may contain tens of billions of planets roughly the size and mass of Earth."

Such "exoplanets" include an unusual one which orbits (contrary to those in *our* solar system) in the opposite direction of the spin of its sun.

Travel to other planets is more promising for science *fiction* than for science. *Time* magazine (6-25-2001): "Even the speediest galactic ark would have to travel hundreds of years, during which multiple generations would live and die on board, before reaching even a nearby star like Proxima

---

* "Carl Sagan calculated the number of probable planets in the universe at as large as 10 billion trillion—a number vastly beyond imagining." —Bill Bryson

Centauri, 4.3 light-years away." (Reduce Earth to about the size of a pea, the *star* would be nearly 10,000 miles distant.) "The best speeds yet achieved by any human object are those of the Voyager 1 and 2 spacecraft, which are now flying away from us at about thirty-five thousand miles an hour," says Bill Bryson.

These twin probes (which have given us fly-by details from Jupiter and Uranus, as well as while passing Neptune) have been sending data for thirty-four years. Voyager 1 is now nearly eleven billion miles away, one day to head out of the solar system and into interstellar space. Signals take more than twelve hours to reach us. But if it were to reach the nearest star within a human lifetime, it would need to be traveling at 10,000 miles per second.

Our sun, which is a hundred times the size of Earth, makes up 99.9% of all the material mass in *our* solar system. And of the billion billion billion tons of the sun's mass, five million tons is being burned as energy every *second*. As a consequence, there will be a (distant) time when the sun dies; in that process, it will expand to engulf Mercury, then probably Venus and Earth, in its death throes.[*]

Writer Bill Bryson on another variable:

> "Without the Moon's steadying influence, the Earth would wobble like a dying top, with goodness knows what consequences for climate and weather. The Moon's steady gravitational influence keeps the Earth spinning at the right speed and angle to

---

[*] Yet, consider how remote the sun is: walk one step; think of *that* as 3,000 miles. The sun would be a twenty-mile hike away.

provide the sort of stability necessary for the long
and successful development of life. This won't go
on forever. The Moon is slipping from our grasp at
a rate of about 1.5 inches a year. In another two
billion years it will have receded so far that it won't
keep us steady...."

Of the eight planets outward from our sun, it is remarkable
how unique each one is. The readiest example is the relative
weight of a one-hundred-pound person: nearly the same on
Saturn and Neptune, but thirty-eight pounds on Mercury
and Mars, and two hundred fifty pounds on Jupiter.

A day on Mars lasts about equal to a day on Earth; but
a day on Venus passes in about 243 earth days, on Jupiter
and Saturn about ten hours.

Mercury completes its annual orbit in 88 days, Neptune
in 165 years.

Mercury's diameter is about three thousand miles across
(less than half of Earth's), Jupiter's is eighty nine thousand
miles.*

While the axis of the other planets is slightly tilted,
Uranus lies on its side (98°); it rolls like a ball, around the
sun, while other planets spin like tops.

The smallest planet, Mercury (a little larger than Earth's
moon), has daily temperature ranges from about 800° F. to
-280° F.; Venus, though nearly twice as far from the sun,
sees temperatures almost a hundred degrees hotter. Saturn's
and Neptune's winds approach one thousand miles an hour.
Mars has the tallest mountains of all, a volcano two and a

* Jupiter's mass equals that of all the planets combined, in our solar system,
with room to spare.

half times higher than Mt. Everest, and a canyon about as wide as the U.S.

While Earth has one moon,[*] Jupiter has 63. There is an *asteroid* that has two moons. As to asteroids, we know of more than ninety thousand. A couple of Mars' moons may have been asteroids. One of Jupiter's moons, the largest in the solar system, would be a planet if it was orbiting the sun.

We have landed probing or robotic craft on Mercury, Venus and Mars. We have a photo of Earth taken from within about thirteen thousand miles of Saturn (showing just a small orb, lit by the Sun). We have close-up pictures of water-worn pebbles on Mars. We have a photo taken thirty-three inches away from a rock of ice on Saturn's moon, Titan (nine billion miles from here).

Mars might have water-borne life forms beneath its crust. In fact, four of the *moons* in our solar system might have sub-surface water (and organisms) as well.

But at least for Mars, a Cornell planetary scientist states, "Mars was a habitable world at some point early in its history." (Just over 400 years ago, Giordano Bruno, a Catholic monk and astronomer, was burned at the stake in Rome for suggesting extra-terrestrial life forms).

The NASA Cassini space probe detected complex organic molecules in the atmosphere of the Saturn moon Titan. A team headed by planetary scientist Sarah Hörst, at the University of Arizona, replicated Titan's environment.

---

[*] Only sixty-nine years after the 1903 Wright Brothers flight, our astronauts drove a four-wheel cart twenty-two miles on Earth's moon.

According to science writer Andrew Grant, "[She] combined cold nitrogen, methane, and traces of carbon monoxide and exposed the mix to microwaves (which simulate the sun's ultraviolet rays) and oxygen (which rains down on Titan from eruptions on the nearby moon Enceladus). The resulting concoction contained amino acids, the fundamental units of proteins, as well as the five chemical bases that constitute DNA and RNA.

"Perhaps the most notable aspect of Hörst's experiment is what she left out: liquid water, which is crucial for terrestrial life but absent from most of the cosmos, including Titan. 'In the right kind of atmosphere, you can have extremely complex chemistry going on *without* water,' she says."

Oxygen is not a common fixture encapsulating other planets. Yet for our form of life it is necessary. Or, was thought to be: deep-sea researchers have located multicellular animals—three new jellyfish-like species less than a half-inch long—in salt water sediment at the bottom of the Mediterranean Sea. Writes Laurie Salerno (*Discover* magazine, Jan./Feb. 2011): "The finding raises the possibility that complex animal life could exist in all kinds of harsh, oxygen-free environments—on Earth and perhaps in other worlds, too."

The difficulty we'll have in establishing whether life exists on other planets is due to our inability to determine specifically what "life" means: we can't assume that it is defined only by what we earthlings know it to be. On other planets, it may have to meet much stricter and subtler standards. Even here on Earth, there are life forms that have surprised biologists, when they were first encountered. According to the book *One Universe*: "The pinhead-sized tardigrade, which lives in moss and mud in roof gutters and the cracks of paving stones, can withstand pressures 6,000 times greater than at sea level, and temperatures from near absolute zero* to 250 degrees Fahrenheit. It also survives complete dehydration, as well as laboratory exposure to a vacuum and to intense X-rays. Some tardigrades have been revived after lying dormant in dried moss in museums for more than 100 years."

And colonies of cyanobacteria exist beneath Antarctic ice. "Researchers liken these conditions to those on Mars. Dormant ancient microbes, and even plants such as moss, can remain preserved in ice, resuming metabolic activity after thousands to *millions* of years."

Discover magazine:

"Phosphorous is a key component of DNA, but late last year a team of NASA scientists announced they had found a bacterium that could use arsenic instead. 'What else can life do that we haven't seen yet?' wondered lead researcher Felisa Wolfe-Simon."

Bill Bryson:

"Scientists in Australia found microbes known as *Thiobacillus concretivorans* that lived in—indeed, could not live without—concentrations of sulfuric acid strong enough to dissolve metal. A species called *Micrococcus radiophilus* was found living happily in

---

* Nearly -460° Fahrenheit.

the waste tanks of nuclear reactors, gorging itself on plutonium and whatever else was there."

Life Extension magazine:

"Most people think radiation is toxic to all living organisms. Not so with a bacterium called *D. radiodurans*, whose ultra-high levels of antioxidants superoxide dismutase (SOD) and catalase enable it to thrive *inside* nuclear reactors. Radiation acutely kills by inflicting free radical damage to life-sustaining cells. Due to its naturally high antioxidant status, *D. radiodurans* can withstand a radiation dose that is 3,000 times greater than what would kill a human."

And this, just in: "Scientists recently discovered a species of bacteria that live entirely on caffeine." (*Discover*, 9-11)

Astrobiologist David Warmflash, M.D. (*Scientific American*, November 2011), reports:

"Planetary scientists have found that rocks from Mars do make their way to Earth; in fact, we estimate that a ton of Martian material strikes our planet every year. Microorganisms might have come along for the ride. The impacts that launched these rocks into Earth-bound trajectories were violent, high-pressure events, but experiments show that certain species would survive."

The Russians are sending a round-trip space probe to the Martian moon Phobos (which will bring back a scoop of soil, in 2014). Along for the experimental ride are ten diverse species of microorganisms, to test the viability of life forms surviving the interplanetary trip. Among them, a bacteria—*Deinococcus radiodurans*, which Warmflash is "quite sure will survive the trip."

Among their company will be tardigrades (defined as "water animals often regarded as primitive arthropods; invertebrates with an exoskeleton, similar to insects"). At about 1½ millimeters long, you'd have to enlarge one by 500 magnifications for it to be as wide as your four fingers. As a space travel candidate, says Warmflash, "They are extremely resistant to radiation, temperature extremes and even the space vacuum."

A 4.5-billion-year-old piece of rock that was once part of the crust of Mars was discovered as a meteorite in Antarctica. It's estimated to have landed there thirteen thousand years ago, having been dislodged by an impact object sixteen million years ago. And, in 1998, a meteorite was found in West Texas containing large halite crystals, similar to salt, with water inside the crystals that may "predate the sun and planets in our solar system." Such meteorites may have seeded our early planet with the forms which gave rise to life.

In Australia in 1969, according to writer Bill Bryson, a fireball meteorite exploded above a town, raining down chunks of carbonaceous chondrite weighing up to twelve pounds (some two hundred pounds of it). It was determined to be 4.5 billion years old,

> "...and it was studded with amino acids—seventy-four types in all, eight of which are involved in the formation of earthly proteins....Get enough

of those crashing into a suitable place—Earth, for instance—and you have the basic elements you need for life."

"If we define spirituality as 'experiencing our true connection to all that exists,' then the new origin story comes closer than any other to helping us fulfill that longing." —Joel Primack

The Big Bang *began* as a potential universe, within a size less that of a single atom[*] (your *initial* genesis). No matter manifested—only hydrogen and helium—for the first half million years, when atoms began to form, with photons then producing transparent light. So, fast-forward from some thirteen billion years ago to just a little more than four and a half billion years ago, when Earth formed. For two and a half billion years, life on Earth was limited to single-cell organisms; multi-cellular life began to appear only about 1.2 billion years ago. Though dinosaurs lived for more than 100 million years, modern humans appeared only less than 200,000 years ago.[†] If the history of Earth was represented by 100 *years*, mankind emerged during just the last three *weeks*.

---

[*] Princeton University theoretical physicist Paul Steinhardt: "The volume of space we observe today was a quadrillionth the size of an atom" at about 10^{-35} second after the Big Bang.
[†] Compare this with the ninety-million year old sea urchin.

Earth Life

According to Michael Wysession, a professor of geophysics:

"Our planet is 23,000 times older than our race of *Homo sapiens*. It's really hard for us to comprehend. There's a writer, John McPhee, who used an excellent analogy. He described the length of an arm as the age of the Earth. If you consider your shoulder as being the start of the Earth, and you consider the end of your finger as being modern day, if you were to take a nail file and very lightly wipe it across the end of your fingertip, you would erase all of human civilization. That's how small a part of the Earth's time we've occupied."

Astronomer Mark Whittle has compared mankind's tenure with the age of the *universe*. Representing the span of time since the Big Bang as a four-story building, the appearance of the Homo species would account for the final millimeter of the flooring. The written history of our species would amount to the width of a human cell.

A comparative human timeline:

2.5 million years ago:	Homo habilis species, first makers of stone tools
2 million years ago:	Homo erectus
1.8 million years ago:	Migration begins out of Africa
1.6 million years ago:	The use of fire
195,000 years ago:	Homo sapiens (us)
72,000 years ago:	Sewn clothing now worn
35,000 years ago:	Cave paintings by Cro-Magnons
10,000 years ago:	Agriculture and villages
5,000 years ago:	Writing developed

Between Homo erectus and Homo sapiens, the species H. heidelbergensis, H. neanderthalensis and H. florsiensis are considered to have existed. So, including H. habilis and H. erectus with the latter three, five out of six of our species have become extinct, with only H. sapiens surviving—so far. And between H. habilis and H. erectus there is believed to have been another species, Homo ergaster. Thus it may be that six out of seven human species have failed to survive.

New discoveries introduce new questions, and—as an article in *Discover* magazine (September 2003) remarks— hardly a month goes by without news of a significant scientific discovery, including those in the fields of archeology and paleontology. And: any information more than a few decades old is probably being quickly rewritten.

Our human species (Homo sapiens) is of a family called hominids, two-legged primates. Evidence indicates that hominids began walking upright approximately five million years ago, so that's where our human traits began to evolve. Hominid brains began to increase in size, from that of about the size of a chimpanzee's, possibly as recently as two million years ago, nearing our present size at least 160,000 years ago. The maturation of a larger brain made children dependent for a longer time, and may have had an impact on the stability of social interrelations.

Shaping a tool from a stone requires a certain amount of imagination, as does the making of a spear; bone-tipped spears may be as recent as fifty thousand years ago. Barbed-bone fishing hooks have been found that are estimated to be ninety thousand years old.

Today, species of life are becoming extinct at a disturbing rate, due to the traits of one species in particular.

One of the questions that new discoveries have not yet shed light on: did this dominating, self-interested species—H. sapiens—account for the demise of some of the five or six other human species that existed before it? Why is our species the only living survivor? And what can that tell us of the potential demise of *this* species?

What follows are just a handful of examples of the intelligence of life which has evolved in our world.

"Not until 1839," says Bill Bryson,[*] "did anyone realize that *all* living matter is cellular." And, then, the idea was "not widely embraced at first."

At some point before birth, you may be a collection of as many as ten-thousand trillion cells:

[*] Bill Bryson's *A Short History of Nearly Everything* (Broadway Books, 2003), at around 550 pages, is an entertaining, informative and amusing introduction to a range of sciences.

"And every one of those cells knows exactly what to do. . . .Blown up to a scale at which atoms were about the size of peas, a cell itself would be a sphere roughly half a mile across, and supported by a complex framework of girders called the cytoskeleton. Within it, millions upon millions of objects—some the size of basketballs, others the size of cars—would whiz about like bullets. There wouldn't be a place you could stand without being pummeled and ripped thousands of times every second from every direction. . . .The proteins are especially lively, spinning, pulsating, and flying into each other up to a billion times a second. Enzymes (themselves a type of protein) dash everywhere, performing up to a thousand tasks a second. Like greatly speeded-up worker ants, they busily build and rebuild molecules, hauling a piece off this one, adding a piece to that one. . . .

"Typically a cell will contain some 20,000 different types of proteins, and of these about 2,000 types will each be represented by at least 50,000 molecules. 'This means,' says Nuland, 'that even if we count only those molecules present in amounts of more than 50,000 each, the total is still a very minimum of 100 million protein molecules in each cell. Such a staggering figure gives some idea of the swarming immensity of biochemical activity within us.' It is all an immensely demanding process. Your heart must pump 75 gallons of blood an hour, 1,800 gallons every day, 657,000 gallons in a year—that's enough to fill four Olympic-sized swimming pools—to keep all those cells freshly oxygenated. (And that's at rest. During exercise the rate can increase as much as sixfold.)...At any given moment, a typical cell in your body will have about one billion ATP* molecules in it; and in two minutes every one of them will have been drained dry and another billion will have taken their place. Every day you produce, and use up, a volume of ATP equivalent to about half your body weight. Feel the warmth of your skin. That's your ATP at work....

"Finally, cells communicate directly with their neighbors to make sure their actions are coordinated.... Indeed, if not told to live—if not given some kind of active instruction from another cell—cells automatically kill themselves. Cells need a lot of reassurance.... Indeed, some organisms that we think of as primitive enjoy a level of cellular organization that makes our own look carelessly pedestrian. Disassemble the cells of a sponge (by passing them through a sieve, for instance), then dump them into a solution, and they will find their way back together and build themselves into a sponge again. You can do this to them over and over, and they will doggedly reassemble because, like you and me and every other living thing, they have one overwhelming impulse: to continue to be.

"Every cell in nature is a thing of wonder. Even the simplest are far beyond the limits of human ingenuity. To build the most basic yeast cell, for example, you would have to miniaturize about the same number of components as are found in a

* A nucleotide present in, and vital to, energy processes in all living cells.

Boeing 777 jetliner and fit them into a sphere just five microns across; then somehow you would have to persuade that sphere to reproduce."

A paramecium is a single-celled organism which feeds on bacteria in water, swimming with hair-like legs called cilia. The numerous cilia form the external extremities of the paramecium's cytoskeleton, and are composed of bundled tiny tubes called microtubules. Being a single cell, it has no room for cells such as neurons (human neurons are themselves single cells, and each has its own cytoskeleton). Thus, like other one-celled organisms (such as amoebas), it has no brain and nervous system. Yet it has enough intelligence to flee from a dangerous threat, and swim around obstructions.

Some species of ants* are dependent upon varieties of trees for their nesting sites. They have been observed to tear apart vines that might kill their host tree, and also to destroy butterfly eggs whose larvae would devour the plant's leaves. They seek to protect the longevity of their host "with no apparent *immediate* benefit to the ants," states ecologist Mark Moffett.

Another ant species has been known to remove its tree's flowers, forcing the plant's energy into growing larger and thus providing more housing space.

Trees sometimes reward their protectors. A particular Costa Rican shrub "secretes sticky white food globules only

---

* There are an estimated ten thousand trillion ants—about a million per person on earth—and they've been around for some 140 million years.

after the favored ant species moves in, then stops producing them if the colony dies out."

Ants are, also, planned victims of parasites, according to ecologist Steve Yanoviak. A form of nematode doesn't resist being eaten; once ingested, it turns the ant's rear end red, like a ripe berry. This seems to act to attract birds, who eat such an ant—and thus carry and spread the nematode's eggs via its feces. Ants feed on the feces, and the cycle continues.

According to National Geographic (October 1999):

"Plants can communicate with each other. Ilya Raskin, a botanist at Rutgers University... demonstrated this in an experiment. Dozens of tobacco plants, chosen because of their strong chemical response to a particular virus, were placed in two airtight chambers. Tubes carried air between the chambers. The scientists injected the plants in one chamber with the virus. Within two days, those infected emitted a volatile chemical into the air, stimulating the plants in the second chamber to produce chemicals in their leaves that protected them against the virus."

The celebrated African gray parrot, Alex, died in 2007, at age 30, unexpectedly—and ended a promising career, as several obituaries noted. He knew about fifty words for objects; could count and recognize numerals up to six; and distinguish objects by color or comparative size: where there was no difference, he would answer, "None."

A team of German researchers, according to *New Scientist* magazine (5-24-2008) trained a Border Collie to recognize the meanings of hundreds of words. "He could go into another room and retrieve an object he had been asked for, and was even able to do so when asked to retrieve an unfamiliar item from a set of objects for which he had already learned names."

Kanzi, a twenty-six-year-old male bonobo (a species of small African chimpanzee), has learned 348 word symbols, including some abstract concepts like "now" and "bad."

His trainer (psychologist Sue Savage-Rumbaugh) created a keyboard displaying a visualized symbol for each word, which Kanzi operates for communicating. In addition, he is said to comprehend three thousand spoken English words, in command sentences (such as "carry the TV outdoors").

According to *Smithsonian* magazine (November 2006):

"Once, Savage-Rumbaugh says, on an outing in a forest by the Georgia State University laboratory where he was raised, Kanzi touched the symbols for 'marshmallow' and 'fire.' Given matches and marshmallows, Kanzi snapped twigs for a fire, lit them with the matches and toasted the marshmallows on a stick...

"She and her colleagues have been testing the bonobos' ability to express their thoughts vocally, rather than by pushing buttons. In one experiment she described to me, she placed Kanzi and Panbanisha, his sister, in separate rooms where they could hear but not see each other. Through lexigrams, Savage-Rumbaugh explained to Kanzi that he would be given yogurt. He was then asked to communicate this information to Panbanisha. 'Kanzi vocalized, then Panbanisha vocalized in return and selected "yogurt" on the keyboard in front of her,' Savage-Rumbaugh tells me."

Nim, a chimpanzee, was taught sign language, from three months of age, by researchers in facilities at Columbia University, starting with such words as "drink," "sweet" and "more." Behavioral psychology students documented Nim signing twenty thousand combinations of words. His cravings included pizza, coffee and cigarettes. According to an article about him,

"Nim became a major attraction for Columbia students. At night, joints were passed around the living room. Sometimes Nim would take a puff and inhale with pleasure....When Nim was given a group of photographs to sort—images of chimps, including himself, mixed up with those of humans— he would put his own picture in with the humans."

He died of a heart attack, about midlife, at age 26. (A co-star of the early Tarzan movies, the chimpanzee Cheeta died at age 80, in December 2011, of kidney failure.)

From a Princeton University text, *Margins of Reality* (Jahn and Dunne):

"Other scientists and philosophers have pondered whether atomic structure may be characterized by its own intrinsic form of consciousness. By the definition of consciousness proposed (by us)—'any functioning entity capable of generating, receiving, or utilizing information'—atoms and molecules would certainly qualify; for they have the capacity to exchange information with each other and with their environment, and to react to these in some quasi-intelligent fashion."[*]

At ten weeks, the human brain is about half an inch long, looking like a large lima bean, but already with two distinct hemispheres, and also neurons. A neuron is a cell of the nervous system, with a nucleus, branching axon (transmitter) and several branched dendrites (receivers).

At twenty weeks, the brain is two inches long, with the basic shape it will retain as it grows.

The neurons are born in a lower part of the brain (the ventricles), divide, and migrate to the cortex, the outer gray matter lying over much of the brain. It is how they migrate that is of interest.

Glial cells grow from the ventricles to the cortex, as long, thin fibers. In the words of a neurosurgeon (*Discover* magazine, August 1998), a neuron "hugs" a radial glial cell,

---

[*] A colony of a particular soil bacteria, "about eight centimeters in diameter, contains 100 times more bacteria than the number of people on earth." According to Tel Aviv University physics professor Eshel Ben-Jacob: "Acting jointly, these tiny organisms can sense the environment, process information, solve problems and make decisions so as to thrive in harsh environments."

like the tendril of a vine, by the neuron's leading part—feeling around as it reaches for the next point to grip in its upward climb, and pulling the cell nucleus behind it. Once in the cortex, it releases its grip on the glial fiber so that another upward-mobile neuron can follow behind it. Early on, the trip takes only a day*; as the brain expands, it begins to take as long as two weeks.

The magazine article says, "There are a hundred billion neurons in the adult human cortex, and all of them got there by migrating" from their original birthplace to the cortex, climbing a radial thread almost like a snail climbs a beanstalk. This proceeds for about three months, during which neurons establish the locations of their axons and dendrites.

Physics professor Brian Greene puts the power of your brain into contemporary perspective:

"The human retina, a thin slab of 100 million neurons that's smaller than a dime and about as thick as a few sheets of paper, is one of the best-studied neuronal clusters. The robotics researcher Hans Moravec has estimated that for a computer-based retinal system to be on a par with that of humans, it would need to execute about a billion operations each second. To scale up from the retina's volume to that of the entire brain requires a factor of roughly 100,000; Moravec suggests that effectively simulating a brain would require a comparable increase in processing power, for a total of about 100 million million (10^{14}) operations per second. Independent estimates, based

---

* With a billion, or more, neurons migrating each *day*.

on the number of synapses in the brain and their typical firing rates, yield processing speeds within a few orders of magnitude of this result, about 10^{17} operations per second. Although it's difficult to be more precise, this gives a sense of the numbers that come into play. The computer I'm now using has a speed that's about a billion operations per second; today's fastest supercomputers have a peak speed of about 10^{15} operations per second (a statistic that no doubt will quickly date this book). If we use the faster estimate for brain speed, we find that a hundred million laptops, or a hundred supercomputers, approach the processing power of a human brain."

"There is something marvelous in the fact that we barely understand what most of the cells in our brains are doing," declares Carl Zimmer (*Discover* magazine, September 2009).

"Now scientists are figuring out how to observe astrocytes in living animals, and learning even more about the cells' abilities....If astrocytes really do process information, that would be a major addition to the brain's computing power. After all, there are many more astrocytes in the brain than there are neurons."

The brain contains about a trillion glial cells, ten times the number of neurons. The *radial* glial cells first provide a scaffolding for the locating of neurons in their position in the brain, but then transform into another type called *astrocytes,*

"...the most abundant of all the cells in the brain....And they speak in a chemical language of their own....They have at least some of the requirements for processing information the way neurons do....They are also the most mysterious. A single astrocyte can wrap its rays around more than a million synapses.* Astrocytes also fuse to each other, building channels through which molecules can shuttle from cell to cell."

The (star-shaped) astrocyte cells have been observed (in the brain stem of laboratory rats) to signal the neurons[†] which influence breathing, when excess carbon dioxide has been inhaled. This caused the rats to breathe more deeply and absorb more oxygen. These glia "are even more sensitive than neurons," remarks one molecular physiologist (*Science* magazine, July 2010).

In a biophysics experiment at Harvard, the subject was a quarter-inch nematode, a type of transparent worm which has approximately a hundred muscles, about three hundred neurons and around five thousand connections among them (*Scientific American* magazine, March 2011).

"Researchers have come a step closer to gaining complete control over a mind, even if that mind is

---

* There are some one hundred trillion synapses, or neural connections, in the human brain—more than a thousand times the number of stars in the Milky Way.

† *Smithsonian* magazine: "By borrowing a gene from fluorescent jellyfish and inserting it into the DNA of worms or mice in the lab, scientists have made neurons glow."

smaller than a grain of sand. A team at Harvard University has built a computerized system to manipulate worms—making them start and stop, giving them the sensation of being touched, and even prompting them to lay eggs—by stimulating their neurons individually with laser light, all while the worms are swimming freely in a Petri dish."

Writer Bill Bryson summarizes the marvel of DNA and genes:

"...DNA—'the most extraordinary molecule on Earth,' as it has been called. DNA exists for just one reason—to create more DNA—and you have a lot of it inside you: about six feet of it squeezed into almost every cell. Each length of DNA comprises some 3.2 billion letters of coding, enough to provide $10^{3,480,000,000}$ possible combinations, 'guaranteed to be unique against all conceivable odds,' in the words of Christian de Duve. That's a lot of possibility—a one followed by more than three billion zeroes. 'It would take more than five thousand average-size books just to print that figure,' notes de Duve. Look at yourself in the mirror and reflect upon the fact that you are beholding ten thousand trillion cells, and that almost every one of them holds two yards of densely compacted DNA, and you begin to appreciate just how much of this stuff you carry around with you. If all your DNA were woven into a single fine strand, there would be enough of it to stretch from the Earth to the Moon and back, not once or twice but again and again. Altogether, according to one calculation, you may have as much as twenty million kilometers of DNA bundled up inside you.

"The simple amoeba, just one cell big and without any ambitions but to exist, contains 400 million bits of genetic information in its DNA—enough, as Carl Sagan noted, to fill eighty books of five hundred pages.

"'Wherever you go in the world, whatever animal, plant, bug, or blob you look at, if it is alive, it will use the same dictionary and know the same code. All life is one,' says Matt Ridley. We are all the result of a single genetic trick handed down from generation to generation nearly four billion years, to such an extent that you can take a fragment of human genetic instruction, patch it into a faulty yeast cell, and the yeast cell will put it to work as if it were its own. In a very real sense, it *is* its own."

Discover magazine:

"If uncoiled, the DNA in all the cells in your body would stretch 10 billion miles—from here to Pluto and back....Aside from bacteria, the smallest genome belongs to the intestinal parasite *Encephalitozoon intestinalis*, with a trifling 2.3 billion base pairs."

David Freedman in *The Atlantic* magazine (July/August 2011):

"If you think genes don't affect how people behave, consider this fact: if you are a carrier of a particular set of genes, the probability that you will commit a violent crime is four times as high as it would be if you lacked those genes. You're three times as likely to commit robbery, five times as likely to commit aggravated assault, eight times as likely to be arrested for murder, and 13 times as likely to be arrested for a sexual offense. The overwhelming majority of prisoners carry these genes; 98.1 percent of death-row inmates do....By the way, as regards that dangerous set of genes, you've probably heard of them. They are summarized as the Y chromosome. If you're a carrier, we call you a male."

Another major player in our body and world is microbial life. Says *Discover* magazine (March 2011): "Two hundred trillion microscopic organisms—bacteria, viruses, and fungi—are swarming inside you right now. The largest collection, weighing as much as four pounds in total, clings to your gut, but your skin also hosts more than a million microbes per square centimeter." Thus, microbes outnumber cells in the body by about twenty to one.*

Further, reports Bill Bryson, "...if you totaled up all the biomass of the planet—every living thing, plants included—microbes would account for at least 80 percent of all there is, perhaps more."

---

* "There are roughly 100 trillion cells in the human body, but of those 100 trillion cells, roughly 90 trillion of them are actually different kinds of bacteria—which means that only 1 out of every 10 celss in your body is actually your own." —Biology professor Bruce Fleury.

In just the past thirty years, we've seen some remarkable developments, including: AIDS first reported; first computer virus released; first artificial heart transplant; a new pure-carbon molecule discovered; ozone hole detected in our stratosphere; Challenger space shuttle explodes, killing crew; Chernobyl nuclear reactor meltdown; interstellar space probe Voyager 2 passes Neptune; Hubble Space Telescope launched; intergovernmental panel warns of climate-warming change; comet observed slamming into Jupiter; FDA approves genetically modified tomato; "top" quark, keystone of subatomic nuclei, discovered; planet detected orbiting a sun-like star; lamb cloned from an adult sheep; Sojourner rover begins Mars exploration; Deep Blue computer beats grandmaster at chess; dark energy identified; adult brains discovered to grow new neurons; West Nile virus enters U.S.; lab-grown bladders implanted in dogs; Wikipedia introduced; water ice detected on Mars; a tsunami devastates Indonesia; Huygens space probe lands on (Saturn moon) Titan; first face transplant performed; skin cells converted to stem cells; Swine Flu pandemic; synthetic bacterium engineered.... Enough, for one generation?*

Yet, with all our technological advances, a cataclysmic event could at any time eradicate the human enterprise.

During the late Cretaceous period,† an asteroid estimated to be six miles wide and traveling hundreds of miles per hour struck Earth in the region of what is now Yucatan. Its explosive impact was in the range of one hundred million million tons of TNT. The portion of the asteroid that landed

* *Discover* magazine, October 2010.

† Considered to be 64.5 million years ago, before the emergence of the first primates.

on the waters of the Gulf of Mexico created waves that were miles high. Dinosaurs in distant locales were knocked off their feet by the ground impact, which left a crater more than thirty-eight miles wide.

Within the first three hundred miles, life was extinguished. Vegetative fires killed more life, over hundreds of thousands of acres. Tidal waves threw oceanic fish and sea vegetation onto land, where more life forms died. Volcanic activity was stimulated in parts of the world, even on the opposite side, where more living forms were engulfed in lava. Smoke and ash limited vision, in wide areas, to a few feet; many more animals died of asphyxiation, and plants died from deprivation of nine months of sunlight. A rotten stench pervaded the atmosphere.

Many remaining life forms starved; some others attacked and ate survivors. Rain caused erosion; and there were two wintry years; the cold and wet conditions killed more life. Four-fifths of the species of animal and plant life succumbed; especially animals of more than fifty pounds, including all non-avian dinosaurs.[*]

A "space object" impacted the Indian Ocean around 2800 B.C. that was upwards of a couple of miles wide, according to an article in *The Atlantic* (June 2008), causing a tidal wave 600 feet high.

In 536 A.D., an object nearly a fifth of a mile around hit the ocean north of Australia at an estimated fifty thousand miles per hour, with an impact estimated to equal a thousand nuclear bombs.

---

[*] "The history of any one part of the Earth, like the life of a soldier, consists of long periods of boredom and short periods of terror." —British geologist Derek Ager

In March 2004, an asteroid nearly a hundred feet across "shot past Earth, not far above the orbit occupied by telecommunications satellites." (Search "2004 FH" at Wikipedia and you can watch it pass through our night sky.)

On November 8, 2011 (reported Associated Press), an asteroid a quarter-mile wide—"bigger than an aircraft carrier"—passed within 202,000 miles of Earth, "just inside the moon's orbit...from the direction of the sun, at 29,000 miles per hour." Named 2005-YU55, astronomers had been tracking it since noticed six years prior; if one that size were to hit the Earth, they calculate, "it would blast out a crater four miles across and 1,700 feet deep," possibly result in a magnitude 7 earthquake, or raise an ocean tsunami seventy feet high.

There are nearly fifty-four hundred "near-Earth" asteroids and comets, 740 of these are a half mile (or more) wide.* NASA considers at least 186 to be "impact risks" (www.neo. jpl.nasa.gov/risk).

A major concern is an asteroid explosion as it enters our atmosphere. Such an event occurred in Siberia in 1908, the object only about a hundred feet wide.

> "The blast had hundreds of times the force of the Hiroshima bomb and devastated an area of several hundred square miles. Had the explosion occurred above London, or Paris, the city would no longer exist....Right now, astronomers are nervously tracking 99942 Apophis, an asteroid with a slight chance of striking Earth in April 2036. Apophis is also small by asteroid standards, perhaps 300

---

* There is at least one asteroid that is 35 miles long (called 243 IDA).

meters across,* but it could hit with about 60,000 times the force of the Hiroshima bomb—enough to destroy an area the size of France."

Citing a researcher, Mark Boslough:

"If, as Boslough thinks, most asteroids and comets explode before reaching the ground, then this is another reason to fear that the conventional thinking seriously underestimates the frequency of space-rock strikes—the small number of craters may be lulling us into complacency....I asked William Ailor, an asteroid specialist at The Aerospace Corporation, a think tank for the Air Force, what he though the risk was. Ailor's answer: a one-in-ten chance per century of a dangerous space-object strike....And as Nathan Myhrvold, the former chief technology officer of Microsoft, put it, 'The odds of a space-object strike during your lifetime may be no more than the odds you will die in a plane crash—but with space rocks, it's like the entire human race is riding on the plane.'"

"Researchers have identified more than 2,300 asteroids and comets that are big enough to cause considerable damage on Earth, and could possibly hit us," says Mark Fischetti in *Scientific American* magazine. As of March 2011, there were 991 asteroids up to 100 meters in diameter that "could destroy a city"; another 1,233 up to ten times that size; and 158 more, bigger than one kilometer, that "could destroy civilization."

---

* Nearly a thousand feet.

An asteroid hit, according to one astrophysicist, is "the top thing on the cosmic list of things to worry about happening."

That's because "fewer than one percent of the projectiles" have been located by us, so far. The manager of NASA's Near Earth Object Program reports that the prospective "population is very large."

A particle physicist was quoted in *The Week* (U.K.; 9-15-2001) as saying that "the universe is perched on a terrible precipice," at present: a chance fluctuation in the cosmic "vacuum" could cause a universal reversal*, in which light would disappear and matter would disintegrate on the atomic level in a spontaneous chain reaction.

This was also referred to in a *New York Times* article (1-21-97), mentioning "the possibility that the universal vacuum—the ubiquitous empty space of the universe—might actually be a false vacuum.

"If that were so, something might cause the present-day universal vacuum to collapse to a different vacuum, of a lower energy. The effect, propagating at the speed of light, would be the annihilation of all matter in the universe. There would be no warning for humankind: the earth and its inhabitants would simply cease to exist at the instant the collapsing vacuum reached the planet."

The editor of *Discover* magazine, Corey Powell, says that a collapse of the cosmic vacuum "probably happened in

* The quantum vacuum is a constant in space and time.

the first 10^{-32} second of the universe's life." What are the prospects of it occurring again? "(Until we) develop a better theory of physics, there is no way to judge its likelihood."

In a *Discover* magazine article (October 2010), Powell wrote a paragraph about another possible cataclysmic scenario.

"Back in the 1970s, when it seemed that the sun was not emitting the expected number of particles known as neutrinos, some solar physicists proposed that our star might go through million-year stretches of reduced activity, during which time its brightness could drop by perhaps 40 percent. Although the initial evidence for a solar shutdown evaporated, the mechanism remains possible. Such a dip would put our planet in a deep freeze, and in fact paleontologists now find evidence of one such episode of extreme cold (nicknamed 'Snowball Earth') about 650 million years ago. If the sun dimmed significantly today, the oceans would gradually freeze solid, and most multicellular life on Earth—humans included—would probably go extinct."

He matched that with another prospect.

"Dark matter is the heavy but invisible (and as yet undetected) stuff believed to hold galaxies together. If a clump passed near our sun, its pull could shake loose comets from the outer solar system, sending some of them crashing into Earth. Australia-based astronomer Kenji Bekki claims that one such passage happened millions of years ago, forming a ring of stars called Gould's Belt. If a dark cloud headed right into (and through) Earth, that might have dire consequences too. In theory, dark matter particles could interact with each other, giving off gamma radiation. Afsar Abbas, a physicist in India, suggests that the radiation would not only cause a wave of mutations but also heat up Earth's interior and trigger massive volcanism, leading to a double extinction. Odds: indeterminate. Dark matter is so elusive that an event could be starting right now and we might not know it."

The sun occasionally spews out, in what is called a coronal mass ejection (CME), billions of tons of an ionized gas, or plasma. Balls of these charged high-energy particles travel through space.* A report of the National Academy of Sciences, according to *New Scientist* magazine (3-21-2009), says that there are now three hundred key transformers throughout the U.S. that could be knocked out by a CME, cutting off power to more than 130 million people in a span of one and a half *minutes*. Space weather stations *might* give us a fifteen-minute warning.

Water supplied by electric pumps will dry up after half a day, as will gasoline that must be pumped. Vehicle traffic will thus stop, including deliveries to supermarkets. Electric and subway trains will be stalled too.

Back-up generators will stop, with lack of fuel; hospital supplies will last no more than three days.

A melted transformer hub cannot be repaired, only replaced. There are not many spare transformers stockpiled.

---

* At 900,000 miles per hour.

And as for sources of power even during repair, electricity is required for natural gas and fuel pipelines. Coal-fired power stations will exhaust their reserve, meanwhile, in about a month. Nuclear power stations are programmed to shut down immediately in the event of serious grid problems, and not allowed to restart until the power grid has been restored.

No power for heating, cooling or refrigeration, will affect the manufacture of medicinals; there are a million people dependent on insulin alone.

A Cornell plasma physicist says the situation "would be like a Hurricane Katrina, but ten times worse!"

Help from Europe, where the electricity grids "are highly interconnected and extremely vulnerable to cascading failures"? The U.S. could be in for a loss of tens of millions of lives, says the magazine. Quoting one authority, "It could conceivably be the worst natural disaster *possible.*"

So, we're not just talking about the computers at your bank being down. We're talking about the electric grid nationwide—hampering any sort of recovery.

The head of NASA's planetary division has said of a CME hitting Earth's magnetic shield: "The Earth can't cope with the plasma" from a solar storm. "The CME just opens up the magnetosphere like a can-opener, and matter squirts in."

The human lives that are dependent upon the power grid, incidentally, are subject to annihilation by more than a CME. David Nichol, a professor of computer engineering and a consultant to the Homeland Security Department, has described how easily "a rogue state or terrorist group" could destroy "critical civilization infrastructure anywhere in the world" (*Scientific American*, July 2011), while "keeping operators from knowing that anything is amiss" until fatally too late.

"...a coordinated cyberattack on multiple points in the grid could damage equipment so extensively that our nation's ability to generate and deliver power would be severely compromised for weeks—perhaps even months....Computers control the grid's mechanical devices at every level, from massive generators fed by fossil fuels or uranium, all the way down to the transmission lines on your street. Most of these computers use common operating systems such as Windows and Linux, which makes them as vulnerable to malware as your desktop PC is."

Nichol says: "According to a 2004 study by researchers at Pennsylvania State University and the National Renewable Energy Laboratory in Golden, Colo., an attack that incapacitated a carefully chosen minority of all transmission substations [as few as eight percent] would trigger a nationwide blackout." And he points out that even "military bases rely on power from the commercial grid."

A sidebar, in the magazine, gave examples of real-life cyberattacks:

January 2003: The Slammer worm bypasses multiple firewalls to infect the operations center at Ohio's Davis-Besse nuclear power plant. The worm spreads from a contractor's computer into the business network, where it jumps to the computers controlling plant operations, crashing multiple safety systems. The plant was off-line at the time.

January 2008: A senior CIA official reveals that hackers have frequently infiltrated electric utilities outside the U.S. and made extortion demands. In at least one case, the hackers were able to shut off the power supply to several (unnamed) cities.

"Malicious software called TDSS," reported *Discover* magazine (September 2011), "has conscripted more than 4.5 million computers" into a consolidated network—"practically indestructible because it can operate without a central command center."

The director of a research institute concerned with computer security, Scott Borg, says of cyberattacks (*Scientific American*, November 2011):

"After lying dormant for months or years, malware could switch on without any action on the part of those who launched it. It could disable emergency services, cause factories to make defective products, blow up refineries and pipelines, poison drinking water, make medical treatments lethal, wreck electric generators, discredit the banking system, ground airplanes, cause trains to collide, and turn our own military equipment against us....The malicious part of the malware might be the sequence of operations that causes a *normal* instruction to be carried out at exactly the *wrong* time....We don't actually know how to scan for malware. We can't stop it, because we can't find it. We can't always recognize it, even if we are looking right at it....The very computers we are using to search for malware might be the vehicles delivering it....
If the first time a malicious program operates it turns your missiles back at you, fries your electric generators or blows up your refineries, it won't do

much good to recognize it by that behavior....We are stopping most malware, most of the time. But we don't have a reliable solution for the cases where it might matter most."

✧

Some say the world will end in fire,
some say in ice.
From what I've tasted of desire
I hold with those who favor fire.

—Robert Frost

According to *Harper's* magazine (November 2007; Mark Fischetti): "Nine countries could kill many people on a moment's notice by launching missiles carrying nuclear warheads. A tenth, Iran, may be weaponizing uranium. The U.S., Russia and China can bomb virtually any country with long-range ballistic missiles and, along with France and the U.K., could do the same using submarines."

The nuclear submarines of Russia, England and the U.S. can "roam almost anywhere," having a navigational range of about five thousand miles, hauling a "full payload."

Russia and the U.S. alone have 11,500 operational warheads "ready to deploy." Stored around the U.S. are another 9,900 warheads (and 15,000 parked in Russia) mainly at (targetable) nuclear-bomb bases, submarine bases and in rural missile silos.

One kiloton equals a *thousand tons* of TNT—the size that fell on Hiroshima in 1945. A megaton is equal to one *million* tons of TNT. The Soviets were reported to have a *fifty megaton* bomb in 1961. "Active weapons in the U.S. arsenal" have a "range of yields"—up to 475 kilotons.

As with an asteroid, you might have anywhere from ten to thirty minutes of warning to finish up whatever you're doing that's important.

Writer Bill Bryson observes that modern humans

"...have existed for only about 0.0001 percent of Earth's history. But surviving for even that little while has required a nearly endless string of good fortune....Of the billions and billions of species of living thing that have existed since the dawn of time, most—99.99 percent—are no longer around. Life on Earth, you see, is not only brief by dismayingly tenuous. It is a curious feature of our existence that we come from a planet that is very good at promoting life, but even better at extinguishing it."

"We shall require a substantially new manner of thinking, if mankind is to survive." —Albert Einstein

Atomic Unreality

"In the experiments about atomic events we have to do with things and facts, the phenomena that are just as real as any phenomena in daily life. *But the atoms or elementary particles themselves are not real;* they form a world of potentialities or possibilities rather than one of things or facts."*

Astrophysicist Bernard Haisch said about light, in a magazine article:

"If you could ride a beam of light as an observer, all of space would shrink to a point, and all of time would collapse to an instant. In the reference frame of light, there is no space and time. If we look up at the Andromeda galaxy in the night sky, we see light that (from our point of view) took *2 million years* to traverse that vast distance of space. But to a beam of light radiating from some star in the Andromeda galaxy, the transmission from its point of origin to our eye was *instantaneous.* There must be a deeper meaning in these physical facts, a deeper truth about the simultaneous *interconnection* of all things."

* Werner Heisenberg

And, in the overall purpose of his article, he had something of more direct interest to describe:

"Any microscopic object will always possess a residual random jiggle, thanks to quantum fluctuations. Radio, television and cellular phones all operate by transmitting or receiving electromagnetic waves. Visible light is the same thing; it is just a higher frequency form of electromagnetic waves. At even higher frequencies, beyond the visible spectrum, you find ultraviolet light, X-rays and gamma-rays. All are electromagnetic waves which are really just different frequencies of light....And if you add up all these ceaseless fluctuations, what you get is a background sea of light whose total energy is enormous: the zero-point field. The 'zero-point' refers to the fact that even though this energy is huge, it is the lowest possible energy state. All other energy is over and above the zero-point state. Take any volume of space and take away everything else—in other words, create a vacuum—and what you are left with is the zero-point field. We can *imagine* a true vacuum, devoid of everything, but the *real-world* quantum vacuum is permeated by the zero-point field with its ceaseless electromagnetic waves....Since it is everywhere, inside and outside of us, permeating every atom in our bodies, we are effectively blind to it. It blinds us to its presence. The world of light that we *do* see is all the rest of the light that is over and above the zero-point field."

Light is described, scientifically, as electromagnetic waves—energy—with perfect velocity. Most of us have a general comprehension of, say, a speed of one hundred miles per hour. Light travels 660,000,000 miles per hour. At some 186,000 miles-per-second, it could circle our globe seven and one-half times in one second.

If you are driving toward me at 50 m.p.h. while I am driving toward you at 30 m.p.h., your speed—*relative* to me—is not actually 50 m.p.h.

Similarly, if you are driving away from me at 50 m.p.h. while I am *following* you at 30 m.p.h., your speed (relative to me) is not actually 50 m.p.h.

If you are driving 50 m.p.h. and I am following you at 50 m.p.h., your speed *relative to me* is zero.

The speed of light is 670 million m.p.h. (rounded). If I were to follow a projected beam of light at 670 million m.p.h. (experiments in Einsteinian relativity have shown) I would *not* keep pace with the light ray; it would *not* appear to be traveling at my speed, nor appear to stand still in relation to me. It would continue to constantly proceed ahead of me at 670 million m.p.h., *regardless* of my speed relative to it. Its speed is defined as "absolute," relative to any other.

Perhaps the most remarkable element, or phenomena, in the physical universe is what we know as light. In quantum terms, this generally reduces down to its *quanta*, the photon. These "particles" of energy have no mass, and (despite being a source of electromagnetic radiation) no charge. Always in motion, they are that which has the capability of traveling at light speed.

While David Bohm (like some others) has suggested that matter is a form of light which has "condensed" at less than light speed, a former University of Pittsburgh professor, Dr. Ernest Sternglass, has concluded that photons can be converted into matter, and matter into photons. Such processes appear to be involved when gamma rays strike our atmosphere, and can be produced in a particle accelerator.

More than any other element, it is light (particularly in its constituent form as photons) which transmits information to us from the cosmos. There are more than a billion photons for every proton or electron (atomic particles).

During the first sixty thousand years after the Big Bang, prior to the organization of matter, photons existed as an opaque fog. For a period, as matter was eventually coalescing, the energy we call light actually outweighed all of atomic matter.

After eighty thousand years, the photon fog became transparent. We are able to see back through the transparent universe, but not able to see back beyond sixty thousand years after the Big Bang because of the early condition of photon fogginess.

The following is not the only such account one can find, in the spiritual literature, of the palpable sensing of the glow of energy within all matter, but it's succinct and verbally clear. Robert Rabbin reports:

> "The sun was just rising above the mountain ridge across the valley. I sat on a concrete planter that surrounded several coconut trees, and fell very silent.

"My head became heavy with silence and my body began to disappear, to dissolve. In another moment there was only breathing, not just my breathing—the respiration of the body—but a breathing of everything around me. I entered that breath of all things and disappeared.

"In this breath was a light, a white light. It emanated from everything. It was everywhere. The leaves and flowers of plants, stone walls, the clumps of dirt, the muddy water, the people beginning to pass by—awareness of breathing and light, yet no perceiver, no body, no self. And tremendous order and intelligence! Such precision and purpose—each thing related exquisitely to the next—everything defined within itself and in relation to everything else, ordered and sustained by the breathing and the light which had no source but was everywhere, streaming, busy and yet *unmoving*.

"This lasted for two days, after which I did not want to talk for a long time.

"The residue of this experience is with me to this day."

In a Los Angeles Times article (1-19-2001), an aspect "of light's intrinsically elusive nature" was reported, of which one scientist was quoted, "Physics doesn't get any more interesting than this." A team led by a Harvard physicist provoked the Times' headline, "Researchers Briefly Bring Light Beam to a Dead Stop."

A pulse of light a half a mile long was directed by a laser into a chamber only a fraction of an inch wide. This forced all of the beam to enter the chamber (traveling at 186,000 miles a second) before any of it reemerged. The light was only delayed for one thousandth of a second before bursting out again at full speed, but it would otherwise have traveled 186 miles in that interval.

In a subsequent experiment, reported by *Discover* magazine, "Physicists at the University of Rochester have coaxed light into traveling backward—and, weirdly enough, to do so faster than light itself."

A pulse of light was beamed through an optical fiber. Writes Alex Stone:

> "Just as one light pulse enters, a second pulse appears at the opposite end, as if by magic. This second pulse then splits in two, with half propagating backward and the other half exiting the fiber. The overall effect is that 'the pulse appears to leave before it enters,' says physicist Robert Boyd, who designed the experiment. No physical laws are violated because the information in the pulse never breaks the light-speed barrier. In recent years, physicists have also learned to slow light or to ramp it up past the usual speed of 186,282 miles per second."

"Empty space is alive and popping with particles and 'virtual' particles that appear and disappear."
—*Sean Carroll, astrophysicist*

"There is general agreement that the quantum vacuum is where everything, that we now know, came from; *even* the matrix of space and time."

So states particle physicist and Oxford professor Frank Close in *The Void* (taking his title from the Rig-Veda). How *relative* space and time, he muses. If someone on a distant star looks at our own star, our sun, she will be receiving this image of light which has traveled space at a time before humans even existed. At the same time, we are seeing its rays as they were originated eight minutes ago.

The research of an atomic physicist, reported in the *New York Times* (1-21-97), confirms that "the vacuum of empty space—devoid of even a single atom of matter—seethes with subtle activity."

"Quantum electrodynamics holds that the all-pervading vacuum continuously spawns particles and waves that spontaneously pop into and out of existence on an almost unimaginably short time scale.

"This churning quantum 'foam,' as some physicists call it, is believed to extend throughout the universe. It fills the empty space within the atoms in human bodies, and reaches the emptiest and most remote regions of the cosmos. In this foam, a typical pair of newborn 'virtual' particles* can survive for only about 10^{-42} second (that is, a fraction of a second, equal to one divided by 10 followed by 42 zeroes)."

---

* "Virtual particles" are not unreal; they have the potential to *become* real.

So, according to this, the "quantum foam" connects the empty space within the atoms in human bodies to the most remote regions of the cosmos.

Cosmologist Brian Swimme, in a magazine interview, discussed Being as emptiness, beyond time, change, and form; the unmanifest, transcendent, Absolute ground of all that is. To him, "it is actually what in physics we call the 'quantum vacuum'" which is constantly producing "elementary particles that then cascade back into nonexistence...every moment of *our* existence is another flaring forth from the quantum vacuum...."

"The atom's electrons go from one state to another state and don't pass through anything in between," no transitional condition: what is known as the "quantum leap."

"David Bohm...says that when you have a particle that is in existence (like an electron), the way it goes from 'here' to 'there' is that it dissolves *into* the unmanifest...and then it reconstitutes elsewhere,"

not necessarily in the same form or particle description.

"The quantum vacuum...is infinitely dense with the possibility of new forms."*

Swimme likens this to personal spiritual transformation, "a death and a rebirth in the form of a new organizing principle of your life, of who you are....These are ancient spiritual ideas now resurfacing within science."†

----

* "In the early universe, one could say the vacuum was fully alive." —Mark Whittle
† Physicist Heinz Pagels: "...what I embody—the principle of life—cannot be destroyed. It is written into the cosmic code..."

80

Scientists (physicists in particular) sometimes have to calculate incredibly short durations of time or miniscule distances in space.

The lower limit on time is called *Planck time*: 10^{-43} seconds, or one 10 million trillion trillion trillionths of a second; written out, 0.(42 zeroes)1.

Because there is a lower limit to a measurable unit of time, we are not able to extrapolate the physical condition (which would have existed in the span of time) up until 10^{-43} seconds following the Big Bang. After that hiatus in time, we are able to determine, for example, that at 10^{-40} seconds after the Big Bang, the universe would have been a millimeter wide in extent.

Similarly, scientists find a lower limit on the extent of space (which is subject to measure), called *Planck length*: twenty powers, or "multiples", less than the width of a subatomic proton.* (It is difficult for us to grasp what these multiples signify. If a millimeter—less than a sixteenth of an inch—represented the figure one million, then one *billion* would be represented as a meter—or about one yard.) Brian Greene: "For a better feel, note that if an *atom* were magnified to be as large as the observable *universe*, the same magnification would make the Planck length the size of an average tree."

---

* "Protons are so small that a little dib of ink like the dot on this *i* can hold something in the region of 500,000,000,000 of them, rather more than the number of seconds contained in half a million years." —Bill Bryson

The quantum leap* in an electron lasts for only about one hundred quintillionth of a second (called an attosecond). That is to one second as one second is to three hundred million years.

But that's not the smallest unit of time, in the context of physics: *that* is so-called Planck time—less than a trillionth of one trillionth of an attosecond (or 10^{-43} seconds).

In a *Discover* magazine article (June 2007), Tim Folger says that's not the half of it: "Time may not exist at the most fundamental level of physical reality."

Time, of course, is relative to start with: "The past, present and future are not absolutes." Einstein too said, "the distinction between past, present, and future is only a stubbornly persistent illusion."

A physicist in France, Carlo Rovelli, points out, "All particles of matter and energy can also be described as waves....An infinite number of [waves] can exist in the same location."[†] So, what is the relevance of time to particles which "could all exist piled together," enveloped in a simultaneous instant?

Time depends upon a subjective observer. Folger: "There is no clock ticking outside the cosmos."

Rovelli: "It may be that the best way to think about quantum reality is to give up the notion of time[‡]—that the fundamental description of the universe must be timeless."[§]

* "If all this damned quantum jumping were really here to stay, I should be sorry I ever got involved with quantum theory." —Erwin Schrödinger
† "A wave is not a material *object* but a *form*." —Physics professor Robert March
‡ "In the Absolute, there is neither time, space, nor cause-and-effect." —Swami Vivekananda
§ "There is no present, no past, no future." —Physicist Fritjof Capra

If there is no time, what does this say about the "eternity" of death—or even the existence of "life," which we presume to have a beginning and end?

The aspect of the observer and the observed being intertwined, in quantum measurements (or "correlations"), "challenges our cherished beliefs in cause and effect," [*] reminds Nobel Prize winning physicist Leon Lederman.

> "One of the more intriguing places where quantum spookiness has arisen is in the very creation of the universe. In the earliest phase of creation, the universe was of subatomic dimensions, and quantum physics applied to the entire universe."

In June 2002, when Tim Folger wrote a *Discover* magazine article about John Wheeler—then nearly age 91; a colleague of Einstein and Bohr; and a physicist for seventy years—Wheeler had participated in the golden age of emerging quantum mechanics. His conclusion, said Folger: we inhabit a cosmos made real in part by our own observations. "A physicist's observations determine...which path [an atom] follows in traveling from one point to another."

Wheeler joined the faculty of Princeton in 1938; in his latest hypothesis, after nearly seven decades of study, "our observations, in the *present*, can affect how a photon behaved in the *past*."

A photon from a distant quasar may have set out (traveling at light speed) even before there was life on Earth. Yet the

---

* "...causality is nothing but a concept, and is not real." —Physics professor Stanley Sobottka

type of experiment which the observing scientist chooses will determine the path the photon takes, according to Wheeler's calculations. Well, more than a calculation: "It has been demonstrated in a laboratory," says Folger, in what is called a "delayed-choice" experiment.

As Stanford physicist Andrei Linde says, "When we look at the universe, the best we can say is that it *looks* as if it were there ten-plus billion years ago." The universe and its observer are a unitary system. "You can say that the universe *is* there only when there is an observer who can say that." A camera could film the universe, but the universe still would not "exist" until a human consciousness acknowledged that what the camera filmed truly exists. Without the universe, *we* are not existent; and the contrary is equally true.

So, as the universe depends on the observer for reification, so too Wheeler suggests does the nature of atomic matter or energy, subject to the wave-function collapse, depend upon what the observer chooses to observe—however far in the past the constituent particles were generated.

A millimeter is .03937 inch, which is smaller than one-sixteenth of an inch as it is marked on a ruler—or, about the width of a numeral one in standard type size (sans serif) printed in boldface. It can also be noted as 10^{-3}, or .001.

A thousandth of a millimeter is called a micrometer (or micron, or 10^{-6} meters); a bacterium or cell, under a microscope, might be a micron in size.[*]

----

[*] "A typical paramecium, for instance, is about two microns wide, 0.002 millimeters, which is really very small. If you wanted to see, with your naked eye, a paramecium swimming in a drop of water, you would have to enlarge the drop until it was some forty feet across. However, if you wanted to see the atoms in the same drop, you would have to make the drop fifteen *miles* across." —Bill Bryson

A nanometer—one thousandth of a micrometer (or 10^{-9} meters)—reaches nearly the limit of size that technicians can work with. One *tenth* of a nanometer (10^{-10}) is about the width of an atom (all atoms are roughly the same size); this is called an angstrom, used in measuring the length of light waves. Think of planet Earth in comparison to an apple (10^{-10}); this is equivalent to an apple in comparison to one atom.*

The orbit of electrons around the nucleus of an atom is about 10^{-10} meters in diameter—so, there is a remarkable amount of space between the core of an atom and the electrons which define its outer limit. Thus the nucleus of the atom is ten thousand times smaller than the atom as a whole (or typically 10^{-14}). If the atom were as wide as a football field—electrons in the end zones—a grape, midfield, would be the nucleus. As a physicist expresses it, "An atom is mostly empty space. Even hard, solid objects [such as your bones] are mostly empty space." (Thus it is estimated that billions of neutrinos can pass through our bodies, in their cosmic travel, every second, unaffected.)

These distinctly-orbiting electrons "sense the presence of other electrons." Consequently there are "interactions of electrons in one atom with electrons in another atom."

The protons and neutrons, which comprise the atom's nucleus, are bonded by what is called the nuclear force, or "strong force." So, particle physics is the study of atomic forces and particles; and the *subatomic* scale involves the principles of *quantum* reality (or "mechanics").

More at the fundament of the nucleus than even its core of protons and neutrons are quarks, which have no

* The Greeks were positing the existence of atoms at about the time of Buddha. But as late as 1800, atoms still were not known to be an existent fact.

internal structure* that is yet known: "We may be looking at a bottom-most layer of the world." What is called the "carrier" of the strong force which holds the quarks together is a particle dubbed the gluon.

A (theoretical) particle field is thought to exist in the atom's empty space between the quarks and electrons; the particles inhabiting the field are known as Higgs bosons. "The Higgs, however, has been established mathematically but not physically. We have no direct evidence of the existence of a Higgs particle."

The Higgs field is believed to exist not only in atoms, but everywhere, even in outer space, as a uniform background. A particle physicist has described it this way:

> "When I walk through the room, then, I'm walking through a sea of the Higgs field. What would be the effect on me? That depends on the interaction of the particles in my body with the Higgs field. One of the things that would happen as I'm walking, if the particles in my body are interacting with the Higgs field, would be some sort of resistance to my motion."

It is thought, in fact, that this as-yet-unverified particle is what provides mass to all particles, including those that comprise your body, and results in inertia.

What is called the electric force holds electrons to the atom's nuclei, and allows atoms to bind into molecules. "The electromagnetic force is responsible for the structure of everyday matter." The force carrier for the electromagnetic force is the photon.†

---

* There is at least one thing that physicists have found to be inseparable: coupled quarks are not divisible.

† There are about 400 million photons per cubic meter (a little larger than a cubic yard) of space. There are a billion photons for every atom.

The carrier for the force of gravity is thought to be a particle named a graviton.

Though physicists regard as real the electrons, protons, neutrons and quarks, these are not objects which can be separately seen, in physical terms.

Physicist Werner Heisenberg: "The atoms, or the elementary particles, are not real; they form a world of *potentialities* and *possibilities*, rather than one of 'things' or 'facts'."

What is "not real" has not even "substance." Physicist Fritjof Capra: "Atoms consist of particles, and these particles are not made of any *material* stuff. When we observe them, we never see any substance; what we observe are dynamic patterns continually changing into one another... interconnections in an inseparable cosmic web....These patterns do not represent probabilities of 'things', but rather probabilities of interconnections."

Capra says, "Subatomic particles...do not exist as isolated entities, but as integral parts of an inseparable network of interactions....To find out what the 'constituents' of these particles are, break them up by banging them together in collision processes (involving high energies)....The resulting fragments are never 'smaller pieces' of the original particles. Two protons, for example, can break up into a great variety of fragments (when they collide at high velocities) but there will never be 'fractions of a proton' among them. The fragments will always be entire...."

And of that debris—those ashes of matter, the raw elements of which your body is comprised—the physicist Joliet-Curie once calculated: the nuclei of all your atoms (which compose 99.9% of an atom's mass) could be packed together into the volume of a minute speck of free-floating dust*—the kind you notice when sunlight streams through the window of a darkened room.

Even then, how unique are "your" raw materials? Physicist Heinz Pagels: "The truth is that the entire material universe, with *all* its variety, is entirely made up out of quantum particles which are completely identical." Physicist Nick Herbert echoes: "All quons, in the same state, are exactly alike....There is no difference whatsoever between electron #123 and electron #137."

In fact, your identity is ultimately even more indistinct than this. Herbert: "The quantum world is not made up of 'objects'. As Heisenberg puts it, 'Atoms are not things'."

Physics professor Robert March: "We cannot understand the universe until we understand the atom."

We routinely speak of matter (or "mass") and energy as two different realities. Physics professor Robert March reminds us, regarding the equation $E=mc^2$: "The formula is sometimes mistakenly referred to as a formula for the conversion of *energy* into *mass*. It is more than that; *it is a statement that, for all practical purposes, the two are identical.* If you want to know how much energy is in a system, measure its mass."

The practical import is that the amount of mass in your hand, for example, could convert into the potential energy of a ten-megaton hydrogen bomb. As another instance, electrons (which have mass) and their anti-matter, positrons, were formed from energy produced by four billion degrees of heat as a consequence of the Big Bang.

Consider that Fritjof Capra is able to declare,

---

* Or, you could pack the entire human population into a sugar cube.

"...we can divide matter again and again, but we never obtain smaller pieces, because we just create particles out of the *energy* involved in the *process*. The subatomic particles are thus destructible and indestructible at the same time."*

Particles collide; annihilate; absorb particles; and emit particles. Matter and energy are simply relative aspects of an undivided wholeness.

The great recent discovery of science is that when things are broken down into their supposed parts, one arrives finally at an irreducible or indivisible element or reality: that which will not accommodate further differentiation. As long as 3,500 years ago, the *Vedas* referred to this as the Imperishable.

The proton and neutron, as the nucleus of an atom, are composed of two or three quarks. While other constituents of the atom are separable,† "it would take an infinite amount of energy to separate quarks," says Professor Richard Wolfson.

Quarks, of course, were not known in Isaac Newton's time. But he said: "It seems probable to me that God in the beginning formed matter in solid, massy, hard, impenetrable, movable particles...even so *very* hard, as never to wear or

---

* "By getting to smaller and smaller units, we do not come to fundamental units, or indivisible units, but we do come to a point where division has no meaning." —Werner Heisenberg
† Though there are eighteen billion tons of force keeping the electron and proton together.

break in pieces; no ordinary power being able to divide what God himself made one in the first creation."

A God that made the quark would have had supreme eyesight. "We think the quark is 10^{-20} times smaller than a proton," says Mark Whittle; something like the size of a bacteria relative to the breadth of our solar system.

Around 400 B.C., Democritus was aware that matter is composed of tiny particles, which he called atoms (meaning, in Greek, "undivided" or "indivisible"). An atom is considered to be a unit (L.: *unitas*, one-ness) of energy; it is mostly empty space, an arena for its smaller particles—protons, neutrons, electrons. If the nucleus (the central part, which constitutes almost all of the mass) of an atom were the size of a marble, its peripheral electrons would be fifty yards from it. One molecule of water contains three atoms (two hydrogen, one oxygen).* If we did not even count the atoms, but just the molecules alone, it would take twenty million years to count the molecules in one *drop* of water—if you were able to count them at the rate of 10,000,000 per second. (And, if we now counted the atoms, we would find that there are one hundred billion billion in that drop of water.) The human body is mostly water, we are told, and water is mostly space in energy.

The dot—"period"—at the end of a printed sentence contains 100 *billion* atoms (of carbon). If you expanded the dot to 110 yards wide, you could see one of these atoms with

* There are molecules in outer space that have as many as thirteen atoms.

the naked eye. If you expanded the dot to 6,215 *miles* wide, you could see the atom's *nucleus* (central core).

The nucleus of a hydrogen atom is a proton. An electron, gyrating around the proton, defines the outer limit of the atom. The electron is a thousand times smaller than the proton. The electron is one ten-millionth the size of the atom as a whole.

The proton contains quarks. To *see* a quark, you would need to expand the dot (or period) to twenty times more distant than the moon (which itself is 238,850 miles away).

Thus, each atom is composed of ("particulate") components which are infinitesimal. An atom is 99.9999999999999 percent empty space. "Its emptiness is profound," says particle physicist Frank Close.

"You" are composed of atoms.

Hydrogen, the lightest known substance, is the most plentiful atom, comprising about seventy-five percent of the universal atom assay. Yet, in terms of mass, in a volume of space compressed to the size of the Earth, the mass of all these atoms would be equivalent to a grain of sand.

To give perspective on the relative volume of an atom, Mark Whittle says that if a typical atom was as large as a marble, your hand (by comparison) would be as big as the Earth.

Bill Bryson gives this comparison: "...one atom is to the width of a millimeter as the thickness of a sheet of paper is to the height of the Empire State Building."

And the atoms in your body, according to Whittle, are older than the Earth, and will outlive the sun.[*]

* "...the average lifetime of a proton is at least ten thousand billion billion billion years." —Paul Davies

Bryson gives these details: "Because they are so long-lived, atoms really get around. Every atom you possess has almost certainly passed through several stars, and been part of millions of organisms on its way to becoming you. We are each so atomically numerous, and so vigorously recycled at death, that a significant number of our atoms—up to a billion for each of us, it has been suggested—probably once belonged to Shakespeare. A billion more each came from Buddha and Genghis Khan and Beethoven, and any other historical figure you care to name."

Whittle also notes, "Atoms are, in a sense [as *you*] actually thinking about themselves."

Atoms, of course, form molecules: many, many.

Bryson says, "At sea level, at a temperature of 32 degrees Fahrenheit, one cubic centimeter of air (that is, a space about the size of a sugar cube) will contain 45 billion billion molecules. And they are in every single cubic centimeter you see around you." This figure is also given (by others) as "one, followed by nineteen zeroes," or "the number of grains of sand in a cubic kilometer" (or about .62 mile on each side).

Look at the palm of your hand: count to three. Some 1,500 trillion of the most common particles in the universe,[*] the neutrino, will have passed through your hand (and on through the globe, out the other side, and beyond). They can travel nine hundred miles in five-thousandths of a second, through solid rock, your brain, or empty space; nothing need be a hindrance to them.

There are hopes of catching samples by scientists in the U.S., Canada, England, Italy, Switzerland, Greece, Russia

---

[*] The Big Bang created as many neutrinos as there are photons of light.

and Antarctica. In 1998, a few thousand of them were snagged by a detector in Japan. (The sun is a major source of neutrinos in our solar environs.)*

When Enrico Fermi submitted a paper on neutrinos to the journal *Nature*, in 1934, it was rejected on the grounds "it contained speculations too remote from reality to be of interest to the reader."

There are three types of neutrinos—electron, muon and tau—and a fourth type may soon be identified. Meanwhile, what has been discovered is that one type (e.g., a muon) can switch its identity to another type (e.g., a tau). As a science writer has commented, "The standard theory of particle physics does not allow that to happen!"

Physics professor Robert March:

"A *fundamental particle*, such as the electron, can be created only if at the same time its own antiparticle is created. Similarly, it can be destroyed only if it encounters one of its own antiparticles. *Field quanta* such as the photon, however, can be freely created or destroyed."

The photon is its own antiparticle. And matter/antimatter annihilation can create photons.

Any particle/antiparticle pair can convert into any other particle/antiparticle pair.

And anti-matter, as well as matter, is affected by gravity.

---

* "Every second the Earth is visited by 10,000 trillion trillion [neutrinos]...
neutrinos do have mass, but not a great deal—about one ten-millionth that of an electron." —Bill Bryson

(Not only matter, but energy too is affected by gravity.)

There are particles, and there are antiparticles, such as a proton and an antiproton. Normally, when the two meet, or collide as in a particle detector, they annihilate each other: +1 and -1 = 0.

A particular short-lived particle, which emerges from particle collisions, is the B meson. Writes Andrew Grant in *Discover* magazine (Jan./Feb. 2011): "During its brief life, this particle rapidly oscillates between matter and antimatter: One moment it's a B meson, the next it's an anti-B meson. This constant wavering should create just as many anti-B mesons as B mesons, but the physicists discovered a clear bias for the matter variety—50.5 percent matter to 49.5 percent antimatter."

If it wasn't for this disparity in particle annihilation—with matter having the edge—there'd be no "universe" as we know it.

Quantum Reality

Early in *Lifetide*, biologist Lyall Watson discusses crystals, the curious, solidified form of a substance in which the atoms or molecules are arranged in a definite pattern that is repeated regularly in three dimensions. Crystals tend to develop shapes bounded by definitely-oriented plane surfaces that are harmonious with their internal structure, sometimes seen in clear, transparent quartz.

"Crystals are vivid examples of the capacity of matter to organize itself. They are regular geometric forms which seem to arise spontaneously, and then to replicate themselves in a stable manner."

Until early in the last century, liquid glycerine was believed to not crystallize. Then, a barrel of glycerine, en route from Vienna to London, crystallized, "due to an unusual combination of movements" in transit. Chemists collected bits from the barrel, and found that these extractions could act as seeds in crystallizing liquid glycerine in their laboratories. Surprisingly, they discovered that although the seed may be applied to one experimental batch of liquid glycerine in their lab, "all the other glycerine in their laboratory began to crystallize spontaneously, despite the fact that some was sealed in airtight containers."

Watson continues: "Clays are extraordinary, layered, crystal structures which have (built into them) what amounts almost to an innate tendency to evolve....Clay has plans." Clays have the ability to absorb other molecules; foreign atoms, such as aluminum, become built in among silica in a molecule of clay. Acquired characteristics can be passed on because of a clay's capacity to (not reproduce, but) replicate. Clay's "memory" lies in its ability to maintain a pattern; such patterns allow some clays, like micas, to induce "ammonium ions and alcohols to solidify into organic components." Certain reactions could give rise to the formation of "membranes and other cell structures. Cell walls could indeed evolve at a later stage...that guard the borders of the modern cell."

"Modern proteins," says Watson, "may have inherited their most-important attribute from ancestral clay." Mother Earth, he says, may be our parent, rather than just our planet.

Further on in the book, Watson gives two accounts that appear to have some relevance to the effect which consciousness has on reality, in its environs: not, in this case, a scientific observer who affects the outcome of a particle physics experiment, but of an effect which is more reminiscent of the across-space interconnection, known as quantum entanglement. (Note that it is not claimed that these experiments have been replicated.) The two accounts follow, as excerpts:

"Helmut Schmidt of Duke University has been involved in several pioneering attempts to track down elusive phenomena. Most of his experiments involve the use of sophisticated electronic apparatus with human subjects, but he has recently tried out one piece of equipment on a cat. Schmidt linked a

96

binary random-number generator in his home to a heat lamp in a garden shed, so that the light turned on and off at strictly-random intervals. When the shed was empty, it did just that, showing no tendency to generate unusual sequences, and keeping the light on exactly half of the time. But when a cat was confined to the shed in cold weather, the machine kept the warm lamp (in the unheated room) on far longer than could be expected according to chance alone."

"At the University of Utrecht, it's the mice that play. Sybo Schouten began by training ten mice to press a lever, in whichever half of their cage an indicator light went on. If the mouse got it right, it received a drop of water as a reward. If it got it wrong, nothing happened. When all the mice were properly trained, Schouten put one in a cage containing lamps but no levers, and another in a cage several *rooms* away with levers but no lamps. Watery rewards appeared simultaneously in both cages if the lighting of the lamp in one, and the pressing of the lever in the other, coincided. The timing of the lamp switch was controlled by a binary random-selector, and the results of the experiment were recorded automatically on punched tape, so that no humans were directly involved.

"In the first series of experiments, several of the mouse pairs consistently produced scores greater than could be accounted for by chance alone. This seems to show that when the lamp lit in the cage of the first mouse, it was able somehow to transmit this information to the second thirsty mouse, who then pressed the appropriate lever to give them both the desired reward."

Watson:

"We are compelled to reexamine all definitions of mind, which see it only as a nebulous entity at the end of a one-way street of sensory traffic....It now becomes necessary for a comprehensive reappraisal of the role of *conception* in structuring reality. Quantum physics already includes consciousness as an essential 'hidden variable' in its equations [though] unlikely to make much difference to the way in which most of us deal with reality on an everyday basis."

He goes on to say:

"I am not necessarily suggesting...that the outer planets didn't exist until we began to look for them. But neither am I prepared to dismiss this *possibility* out of hand....All the methods of detection are manmade. In detecting, we may be creating that which we seek to find."

Watson continues:

"Cyril Hinshelwood, a Nobel laureate in physical chemistry, has suggested that a more-appropriate name for the particles might be 'manifestations.' That sounds right. In purely physical terms, they

have little reality...they appear and disappear....
Perhaps they exist only in consciousness....The
desire for conviction produces its own data, its own
manifestations...."

He cites the Zen students asking the master if it's the flag
or the wind that moves. The reply, "It is your mind that
moves."

Watson:

> "What we regard as ordinary physical matter is
> simply an idea that occupies a world-frame common
> to all minds. The universe is literally a collective
> thought...."

When we think of "particles" or "waves," we are thinking
(or imagining) in macroscopic terms—similar to the way
that we envision an electron "orbit" an atom's nucleus, like
a moon in regard to a planet.

We speak of a particle having "wave-particle duality," yet
there is no duality in that there is no innate *differentiation*
in these descriptive conditions insofar as the particle itself is
concerned. The dualities are in our mind, our imagination.
As a physics professor put it, in an article in *The Sciences*
magazine, "A quanton [nee particle] is not a wave *or* a
particle, but both and neither....We physicists simply lack
that intuition [or, imaginative description] for the in-between
cases."

Physics professor Robert March (*Physics for Poets*):

> "Wave motion is not a mechanical phenomenon,
> because a wave is not a material *object* but a
> *form*....We can have a wave without any movement

of *matter* at all.....Two waves can pass through each other (on a medium) without changing their form."

✧

Professor Brian Josephson of Cambridge University, winner of a Nobel Prize in physics in 1973, has witnessed quantum effects on a scale big enough to *see.*

"Quantum is incomprehensible even to scientists[*].... We don't have a clear *mental* picture of what is going on....You could say the theory is not completely *logical*....You might say the universe is a lot more subtle than we thought....Nature is not just lumps of matter, it's some kind of energy pattern.... Certainly things are not made up out of *particles*.... We do have some sense that 'observation' might help to *construct* reality, and that comes close to the idea that *thought* is involved in the nature of reality.... Some features of mysticism can be connected quite well with properties discovered by science; but I think mysticism goes *beyond* science. I believe aspects of nature, deeper than those discovered by science, are understood in mysticism....The methods of science have so far failed to grasp the subtleties in the way mystical experience has."[†]

Physicist Fritjof Capra (*The Tao of Physics*):

"...the *constituents* of atoms—the subatomic particles—are dynamic *patterns*, which do not exist as isolated entities but as integral parts of

---

[*] "For those who are not shocked when they first come across quantum theory: they cannot possibly have understood it." —Niels Bohr
[†] From an interview with Josephson published in *In Search of the Dead*, Jeffrey Iverson.

an inseparable network of interactions....For the Eastern mystic, all things and events perceived by the senses are interrelated, connected, and are but different *aspects* or manifestations of the same ultimate reality. Our tendency to divide the perceived world into individual and separate things, and to experience ourselves as isolated egos in this world, is seen as an illusion which comes from our measuring-and-categorizing mentality. It is called *avidya*, or ignorance, in Buddhist philosophy and is seen as the state of a disturbed mind...a basic oneness is also the most important common characteristic of Eastern worldviews. One could say it is the very essence of those views, as it is of all mystical traditions. All things are seen as interdependent, inseparable, and as transient *patterns* of the same ultimate reality...."

Physicist/astronomer David Darling, in *Zen Physics*, treats (among other issues) the "intimate connection between the mind of conscious observers and the bringing into being of what is real."

Both (subatomic) energy particles (such as the photons of light) and matter particles (such as electrons—a constituent of all atoms) exhibit both *wave* and *particle* properties; in other words diffuse-like *and* point-like conditions "superimposed" on a fundamental condition: atomic reality is conditional, or relative.

And neither energy nor matter can be accelerated to greater than the speed of light,* at least in our region of the universe.†

A particle's "wave function" tells us that it does not reside at some particular point in space during the moments when it is not being "observed," or (in scientific jargon, "measured"). To this extent, it can be said that a particle does not specifically "exist" when it is not observed: "They have no independent, enduring reality."‡ It is not just that we don't *know where* they are. Eminent physicist John Wheeler has stated that this quantum principle "destroys the concept of the world as just 'sitting out there'....In some strange sense, the universe is a participatory universe."

As a consequence of a conscious action, observation, particles of energy and matter are evoked from a condition of potentiality or possibility to states of tangible materiality and its subsequent events. The phenomenon is called by scientists "wave function collapse," the collapse in which the expectation of our observation becomes an actuality. When you set out to measure, with your laboratory equipment, a particle as a wave, it appears as a wave; were your intention to measure a particle as a particle, it would be present in that form.

> "This is a staggering conclusion...when one remembers that all of the material universe is comprised of subatomic particles!...our most fundamental branch of science implies (the) world cannot even be said to *exist* outside of the subjective act of observation."§

* The speed of light is not to be confused with the frequency of light waves, which is in the range of ten trillion oscillations per second.
† Once operating fully, the Large Hadron Collider is expected to accelerate protons to 99.9999991% of the speed of light.
‡ *Zen Physics*, David Darling.
§ Ibid.

Rainer Maria Rilke makes this same point in a poem:

I know that nothing has ever been real
without my beholding it.
All becoming has needed me.
My looking ripens things
and they come towarwd me, to meet and be met.

Our desire, as Darling describes, to determine reality based on a dualistic choice—this/that, either/or—"actually influences reality in a most fundamental way...our [conscious] intervention fragments the continuous wavelike [indeterminate] nature of the world into *separate*, discrete particles" of matter or energy, thing or event.

"...we break our surroundings down into isolated objects" at the subatomic scale; "a dualistic split from the normal, ongoing state of continuity to a transient state of *individualism*."

Darling goes farther:

"...not only is observership [subjective participation] a mandatory requirement for making reality tangible, but every component of the universe—down to the level of each subatomic particle—is in some peculiar sense immediately 'aware' of what is going on around it."

An experiment can be set up to begin its determination of a particle as a particle, but change the intent (mid-experiment) to determining its wavelike property—in which case, experiment has shown, the particle will accommodate the intent of the "change-of-mind" of the observer. In other words, a photon "somehow knew what [preference, or choice] lay ahead."

The sense of a quanton "being aware of what is going on around it" is reflected in the physics term "nonlocality." Darling: "Nonlocality amounts to zero-delay [instant] communication between two particles, no matter what their separation in distance."

Cause an atom to emit two separate photons, at the same time but in opposite directions. If the electric field of one of them is vibrating in an "up" polarization, the other— by the nature of known physical properties—will always be polarized in a "down" configuration (or vice versa). But until a measurement is performed (if any), the state of polarization of either quanton is *undefined* (not simply unclear at this point). Measure (or observe) the polarized state of either of the quantons (wave-function collapse)[*] and the (alternate) state of the other quanton is instantaneously a determined fact, or actuality—"irrespective of the distance between the particles"; this effect is "real and inescapable." Faster than any energetic signal could travel between the pair—that is, faster than the upward communicable limit of the speed of light—the pair has an immediate awareness of each other's manifested condition, enough so to present a determined state of being.[†]

It is mind that "thereby makes matter real....The universe [from which our minds originate] creates itself, out of itself, *moment by moment.*" Darling adds: "'Subject' [mind] and 'object' [matter] cannot be treated apart....."

---

[*] The collapsed polarization—up or down—will be random. (Mark Whittle: "Quantum fluctuation is inherently random: not even *it* knows what it will do next.")

[†] Experiments, in which "nonlocality" was proven, were suggested in a proposal called Bell's Theorem; the interconnectedness of the particles is described by the word "entanglement."

Motion and rest, energy and mass, time and space are all relative, science tells us:

> "the world cannot be accurately viewed as a complex of distinct *things*....Nothing stands apart.... Incredibly, modern physics—which is the most advanced product of our dualistic way of thinking— has shown that dualism is no longer tenable."

Ironically, it is the "reductionist"—divisive—nature of science which has itself led to such a discovery: through *it*, "we have found that reality has no boundaries."

Scientists at the newly-built Large Hadron Collider, near Geneva, noticed (during the first six months of operation) something that caused the need for "convincing ourselves that what we were seeing was real." According to *Scientific American* magazine (February 2011), "some of the particles created by their proton collisions appeared to be synchronizing their flight paths, like flocks of birds."

At the new facility, the particles are being studied "with higher spatial and time resolution than ever before," and the proton is "one of the most common particles in our universe; and one which scientists thought they understood well." Yet, this finding indicates "the particles may have more *interconnections* than scientists had realized."

Following are various quotations from four physicists, demonstrating a growing awareness that "reality has no boundaries."

Brian Greene:

"If there was any doubt at the turn of the twentieth century, by the turn of the twenty-first, it was a foregone conclusion: when it comes to revealing the true nature of reality, common experience is deceptive. . . .What we've found has already required sweeping changes to our picture of the cosmos. Through physical insight and mathematical rigor, guided and confirmed by experimentation and observation, we've established that space, time, matter, and energy engage a behavioral repertoire unlike anything any of us have ever directly witnessed."

Fritjof Capra:

"Both concepts—that of *empty* space and that of *solid* material bodies—are deeply ingrained in our habits of thought, so it is extremely difficult for us to imagine a physical reality where they do not apply. And yet, this is precisely what modern physics forces us to do. . .space and time are constructs of the mind. The Eastern mystics treated them like all other intellectual concepts; as relative, limited, and illusory."

Vlatko Vedral, theoretical physicist (who ends his book *Decoding Reality* by quoting the Tao Te Ching):

"Quantum physics is indeed very much in agreement with Buddistic emptiness. . . .Everything that exists, exists by convention and labelling and is therefore dependent on other things. So, Buddhists would say that *their* highest goal—realizing emptiness—simply means that we realize how inter-related things fundamentally are. Exactly the same is true in other Eastern religions. Less well known in the West is Advaita Vedanta—a Hindu philosophy that emphasizes the total oneness of the Universe. In this view our perceptions of separate entities is just an illusion—Maya. Even the Universe as a whole only exists by labelling, and not by itself....In quantum physics, as we have argued, particles exist and don't exist at the same time. Here I don't just mean that they exist in different places. I mean that, even in one place, a particle can exist and not exist simultaneously....[The Cappadocian Fathers of the fourth century] proclaimed that, while they believed in God, they did not believe that God exists."

Niels Bohr is reported to have said, "A shallow truth is a statement whose opposite is false; a deep truth is a statement whose opposite is also a deep truth."

A physicist whose books have been translated into ten languages, Brian Clegg tells in *The God Effect* of an entanglement experiment in 1999, using *three* photons rather than just two, carried out by Austrian quantum expert Anton Zeilinger and his team. Then, late in 2002, Denmark's Eugene Polzik and his co-workers entangled two *clouds* of cesium, a metallic chemical element—"each containing *billions* of atoms, in effect an *object*...big enough to be visible to the naked eye!"

After 2003, a study (in the physics department at the University of Chicago) of the magnetic properties of a lithium salt (another metallic chemical element, used in thermonuclear explosives) found that the atoms, which act as tiny magnets, were lined up to create a stronger magnetism than would be expected, evidently revealing that the atoms were in an entangled state. "It seems that quantum entanglements can influence," says Clegg,

> "...a whole magnetic *structure*...something that could be touched and picked up, not an incredibly tiny particle...."[*]

> "What's more...other properties of the salt, including its heat capacity, were influenced by entanglement...and even a small amount of entanglement can produce significant effects in the human-scale world—the 'real' world of tangible physical objects!"

When the gas helium (one of the earliest of elements to form) is cooled to nearly the most extreme temperature (about -460 degrees Fahrenheit) it becomes a "superfluid," with no viscosity at all. Viscosity is "resistance to flow," high in molasses, lower in water. Says one physicist, "If you were to flap your hand around in a superfluid, it would be like flapping your hand in a vacuum. It's like there's nothing there." There is no resistance at all to physical motion. "If you start a ring of superfluid spinning, it will go on spinning forever; there is no friction to stop it," Clegg states. "Most famously, superfluids will spookily attempt to *climb* out of *containers*, as there is no friction to resist the random motion of the molecules."

....

* "Physicists have managed to entangle the quantum states" of "two different squares of synthetically-produced diamond," reports *Scientific American* (February 2012), each about an eigth inch wide, and separated nearly six inches apart.

When some metals, such as aluminum or mercury, are cooled to similar temperatures, they have no electrical resistance when transmitting current, thus are called superconductors. "Place a lightweight magnet above a superconductor, and the magnet will levitate—floating in space." The magnet causes the material to generate its own magnetic field.

Such startling discoveries are further evidence of quantum mechanical effects.

"Entanglement...even now troubles many scientists," notes Clegg. It "seems just as odd to physicists as it does to the rest of us." [Its] "unsettling omnipresence" is what caused Einstein's famous criticism of the very *idea*, Clegg reminds, when he considered it to be "spooky action at a distance," which one might remark of voo-doo. Einstein emphasized that the only reality, for him, was "a world which objectively exists."*

In Walter Isaacson's biography of Albert Einstein, he is quoted, "behind all the discernible laws and connections, there remains something subtle, intangible and inexplicable... beyond anything that we can comprehend."

Does this have anything to do with our thoughts or feelings, and hence our behavior? "Human beings in their thinking, feeling and acting are not free, but are as causally bound as the stars in their motions."

Does this mean that there is no free will? "Everything is determined, the beginning as well as the end, by forces over which we have no control. It is determined for the

* Einstein's long-time friend, nuclear physicist Max Born, "believed that Einstein 'could no longer take in certain new ideas in physics which contradicted his own firmly held philosophical convictions.'" —Physicist Manjit Kumar

insect, as well as for the star. Human beings, vegetables, or cosmic dust, we all dance to a mysterious tune intoned in the distance by an invisible player."

Says Isaacson, "Einstein...forced us to change the way we think about nature"; for starters, time, space and motion. "Quantum mechanics does something similar."

Einstein had great difficulty relating to some aspects of quantum theory, which developed after his groundbreaking work. That a particle could be in a superposition of two potential states simultaneously, he considered as realistic as a pile of gunpowder being at the same moment "exploded and not-exploded."* Even until his death, he did not accept the proposition which later experiments proved. As Isaacson states it,

> "...the timing of the emission of a particle from a decaying nucleus is indeterminate until it is actually observed. In the quantum world, a nucleus is in a 'superposition,' meaning it exists simultaneously as being *decayed* and *undecayed* until it is observed, at which point its wave function collapses and it becomes either one or the other. This may be conceivable for the microscopic quantum realm, but it is baffling when one imagines the intersection between the quantum realm and our observable everyday world."

By the end of 2005, Cornell physicist N. David Mermin was referring to the counterintuitive behavior in the quantum world as "the closest thing we have to magic." In 2006, *New Scientist* magazine reported, "A simple semiconductor chip has been used to generate pairs of entangled photons."

* Yet Einstein is reported to have said, of his theory of relativity, that he did not arrive at "these fundamental laws of the universe through my rational mind."

In entanglement, though two particles may be separated by "billions of miles," affirms Isaacson, they "remain part of the same physical *entity*," so "there is no traditional cause-and-effect *relationship*."

It remains a paradox to many that Einstein perceived "a mysterious tune intoned in the distance by an invisible player," and yet could never accept the extent to which that tune might be being intoned on the level of physics.

Einstein's position was that "whatever we regard as existing (real) should somehow be localized in time and space," as opposed to a particle being merely a probability; or a wave spread out, in principle, through the whole universe. The indeterminate nature of a particle must be a consequence, it was suggested, of some extraneous information which was not yet known, "hidden variables."

A couple of quotations of Einstein (in Corey Powell's *God in the Equation*) are of interest, in the context of his dismay over entanglement:

> "When I am judging a theory, I ask myself whether, if I were God, I would have arranged the world in such a way.... If this Being is omnipotent, then every human action, every human thought, and every human feeling and aspiration is also His work...."

Einstein's theory of general relativity has been affirmed, comments astronomer Hugh Ross,

> "...to better than a trillionth of a percent precision. And even stronger evidence exists for special relativity; it has been affirmed to a precision of better than a ten millionth of a trillionth percent."

Says Brian Greene of quantum physics:

"In the more than eighty years since these ideas were developed, there has not been a single verifiable experiment or astrophysical observation whose results conflict with quantum mechanical predictions.... Like floodwaters slowly rising from your basement, rushing into your living room, and threatening to engulf your attic, the mathematics of quantum mechanics has steadily spilled beyond the atomic domain and has succeeded on ever-larger scales."

Physicist Nick Herbert, who likes to remind that quantum theory "has never made a false prediction," states in *Faster Than Light*:

"Bell's theorem shows that the quantum connection is not a mere theoretical artifact, but corresponds to a real, superluminal link that *actually exists* between any two phase-entangled systems."

This means "not merely that superluminal [faster than the speed of light] connections are possible, but that they are necessary to make our kind of universe work."

It's now less than three hundred years since Sir Isaac Newton opined: "That one body may act upon another at a distance through a vacuum without the mediation of anything else...is to me so great an absurdity, that I believe no man, who has in philosophical matters a competent faculty for thinking, can ever fall into."

In correspondence with Nick Herbert, he sent me this poem:

Quantum Reality
Shall I look at Her
Or shall I not?

Hard, small, separated
If I look;
Soft, spread-out, connected
If I don't.

Hard particle and soft wave: both?
Small right-here and spread-out everywhere: both?
Deep connected yet lonely separate?

Honey
Someday You gotta show me
How You do that.

We think of ourselves as the act-or: the scientist is the "cause," and the measurement which results in a wave-function collapse is the "effect." States physicist David Peat, in *Einstein's Moon,*

> "For two hundred years, physicists have been searching for processes, mechanisms, and *causes* within the world around them. Now quantum theory is saying that, at the level of quantum processes, no such hidden 'causes' exist...."

If two entangled particles must be considered as a single "system," is not the experimenter who determines their spin state also enveloped in that unified system?
Peat says:

113

"At the moment of observation [measurement], the observer and observed make a single, unified whole....Each time we attempt to observe [an electron], we become linked to it so that we can no longer say which is us and which is the atom."

So, where then is the "actor" and the "acted upon"?

In *The Ghost in the Atom,* Paul Davies and Julian Brown mean by "ghost" something like "spirit," or something more like "mind."

"The key role that observations play in quantum physics inevitably leads to questions about the nature of mind and consciousness, and their relationship with matter. The fact that, once an observation has been made on a quantum system, its state (*wave function*) will generally change abruptly sounds akin to the idea of 'mind over matter'. It is as though the altered mental state of the experimenter, when first aware of the result of the measurement, somehow feeds back into the laboratory apparatus, and thence into the quantum system, to alter *its* state too. In short, the physical state acts to alter the mental state, *and* the mental state reacts back on the physical state."

Brian Clegg, in *Before the Big Bang,* writes at length of the contribution of David Bohm to theoretical physics, whose field of expertise was quantum mechanics.

"Bohm got his doctorate under bizarre circumstances at the University of California, Berkeley, during the Second World War. Because he had left-leaning political interests, he was not allowed to join the Manhattan Project to work on the atomic bomb with many of his colleagues. However, his doctoral dissertation covered a subject of significant use to the Manhattan Project, so it was immediately classified and he wasn't allowed to present it or to receive his doctorate. Luckily, Robert Oppenheimer, who headed up the Manhattan Project, had been Bohm's supervisor and was able to get Berkeley to accept that the dissertation was a success, without it ever being officially read."

Though the major part of the Pennsylvania-born Bohm's career was as a professor at the University of London, he earlier taught at Princeton, where one of his colleagues was Albert Einstein, with whom he discussed quantum theory. There are a couple of quotes of Einstein that probably could have been written by Bohm as well:

"The most beautiful emotion we can experience is the mystical. It is the power of all true art and science. He to whom this emotion is a stranger is as good as dead. To know that what is impenetrable to us really exists, manifesting itself as the highest wisdom and the most radiant beauty, which our dull faculties can comprehend only in their most primitive forms—this knowledge, this feeling, is at the center [of] true religiousness. In this sense, and in this sense only, I belong to the ranks of devoutly religious men....What humanity owes to personalities like Buddha, Moses, and Jesus ranks for me higher than all the achievements of the enquiring and constructive mind."

Bell's Theorem, and the experiments it has continued to engender, represents the scientific paradigm-shift of the modern age. While an assistant professor at Princeton in 1951, Bohm suggested a simplified manner in which the proposition of quantum nonlocality (to be known as entanglement) could be experimentally tested. It was this proposal which John Bell, in 1964, developed in a practical form, upon which successful experiments where subsequently performed.

Bohm was interviewed by Rutgers philosophy professor Renée Weber, author of *Dialogues with Scientists and Sages,* in which his description of a quantum field or vacuum suggested to her "the void of Buddhism, the Abyss ['primeval void'] of Christian mystics."

"In nonmanifest reality, it's all interpenetrating, interconnected, one," Bohm said. "Forms may develop out of that which is beyond form." And, "In my view, the implications of physics seem to be that nature is so subtle that it could be almost alive or intelligent." He conceded, "In this definition, it begins to overlap with the area the mystics are interested in."

> Weber: "It will sound to people as if this is a description of religion—that we are constantly grounded in something 'infinite.' Where does it differ from what the great mystics have said?"

> Bohm: "I don't know that there's necessarily any difference."

John Briggs has also written in detail about David Bohm's views. In 1959, Bohm perused a book by spiritual teacher Jiddu Krishnamurti, who said that "the cosmos has no fundamental divisions." More specifically, Krishnamurti often stated (as had ancient spiritual texts), "The observer is the observed." Since Krishnamurti visited England on a regular basis, Bohm arranged to talk with him. Their friendship lasted until Krishnamurti died in 1986 (six years before Bohm); an example of their dialogues can be found in the book *The Ending of Time,* published in 1983.

Briggs says of Bohm's quantum view: "Consciousness is woven implicitly into all matter, and matter is woven out of consciousness." Thus, an individual's consciousness is interwoven with a holistic consciousness.

As Bohm himself came to say, "In quantum experiments, we find that the observer *is* the observed." What connects all objects into one system, Bohm suggested, is

"...a whole field...a field of active information.... Information guides activity....

"That's what I mean when I say there's a basic mindlike quality to particles....There is mind even down at the quantum level." *

---

* "What would it do to our self-definition if we were to become convinced that we have always been part of a whole and are not separate from that which is 'other' than ourselves?...The same dust that makes up the stars of our universe constitutes the substance of our human bodies and perhaps our minds. In fact, we now know that all matter within our universe, from the farthest star to the content of your body and mine, is interconnected." —Retired Episcopal Bishop John Shelby Spong

Bohm and colleague Basil Hiley, in 1975, published a technical paper in *Foundations of Physics*, "On the Intuitive Understanding of Nonlocality, as Implied by Quantum Theory." The point of it (in sixteen pages) was to call associates' attention to entanglement's stimulation of "the radically new notion of unbroken wholeness of the entire universe," and "the challenge of understanding what all this *means.*"

Bohm spelled out what this new discovery means, in a separate writing that was less technical than the paper cited above: "In considering the relationship between the finite and the infinite [or you could read, the relative and the Absolute]...the finite is inherently limited, in that it has no independent existence. It has the *appearance* of independent existence; but that appearance is merely the result of... thought. We can see this dependent nature of the finite from the fact that every finite thing is transient." In other words, the relative appears from, and disappears into, the non-transient Absolute; as the particle appears from and dissolves into the quantum field, or vacuum.

"Our ordinary [relative-oriented] view holds that the... finite is all that there is. But if the finite has no independent existence, it cannot be all that there is." Forms are dependent upon the formless ground of being, for their arising.

"We are, in this way, led to propose that the true ground of all being is the infinite, the unlimited; and that the infinite includes and contains the finite." The unlimited ground of being is all-encompassing, all-inclusive of any limited, impermanent forms.

"The [relative] finite is all that we can see, hear, touch, remember and describe: basically, that which is manifest, or tangible." Matter, energy, time, space, causation, and all other conceptions are in this category of forms that have limitation.

"The essential quality of the infinite [or Absolute], by contrast, is...its intangibility. This quality... suggests an invisible but pervasive energy [a dynamic presence], to which the manifest world of the finite responds [because it *infuses* all]...and this is never born and never dies."

"Quantum theory taxes our very concept of what constitutes science," according to mathematician Amir Aczel in his book *Entanglement.*

"And it taxes our very idea of what constitutes reality....To understand (or even simply accept) the validity of entanglement, and other associated quantum phenomena, we must first admit that our *conceptions* of reality about the universe are inadequate....

"No longer do we speak about 'here *or* there', in the quantum world we speak about 'here *and* there'...."

As in the two-slit experiment,

"...an electron, a neutron, or even an *atom*, when faced with a barrier with two slits in it, will go through *both* of them *at once.** Notions of causality, and of the impossibility of being at several locations at the same time*, are shattered by the quantum theory."

---

* Physics professor Brian Greene: "The double-slit experiment leads us inescapably to a conclusion hard to fathom. Regardless of which slit it passes through, each individual electron somehow 'knows' about both."

Cause-and-effect thinking limits us, and mystics say that we must transcend such dualities.

"Understanding what really happens inside...a quantum system may be beyond the powers of human beings."

Brian Greene, a professor of physics and mathematics at Columbia University, received his doctorate from Oxford University. "Breakthroughs in physics have forced, and continue to force, dramatic revisions to our conception of the cosmos," he notes in *The Fabric of the Cosmos*. Even to a physicist (or, perhaps, especially) it is "astounding" that "two objects can be far apart in space, but...it's as if they're a *single* entity," challenging the "worldview many of us hold.... Quantum mechanics shatters our own personal, *individual* conception of reality." An added twist to entanglement is that its long-distance links are "fundamentally beyond our ability to *control*." Particles,

"...like one of the countless number that make up you and me...[act] as if they are right on top of each other....As a concrete example, if you are wearing a pair of sunglasses, quantum mechanics shows that there is a 50-50 chance that a particular photon—like one that is reflected toward you from the surface of a lake, or from an asphalt roadway— will make it through your glare-reducing polarized lenses: when the photon hits the glass, it randomly 'chooses' between reflecting back or passing through. The astounding thing is that such a photon can have a partner photon that has sped miles away in the opposite direction and yet, when confronted with the same 50-50 probability of passing through another polarized sunglass lens, will somehow do whatever the *initial* photon does. *Even though each outcome is determined randomly, and even though the photons are far apart in space, if one photon passes through, so will the other."*

Such findings present "a frontal assault on our basic beliefs as to what constitutes reality." Before we measure an electron's position, it does not have a definite position, until "the moment we 'look' at it"; it isn't "that we don't *know* the position...the act of measurement is deeply enmeshed in *creating* the very reality it is measuring!"

In regard to the discovery of quantum entanglement, Greene says, "This is an earth-shattering result. This is the kind of result that should take your breath away." Reality operates in unbroken wholeness. The detection of an entangled photon's "spin" causes its partner photon (even at tested distances of more than six and a half miles)

> "...to snap out of the haze of 'probability' and take
> on a definite spin value...that precisely matches...
> its distant companion. And that boggles the mind!...
> They are part of one physical system...parts of one
> physical entity...but do *not* stand in a traditional
> cause-and-effect relationship....No matter what
> holistic words one uses...[they] stay sufficiently 'in
> touch'....What unknown mechanism *enforces* this
> with such spectacular efficiency?"

Any less bizarre universe "may exist in the mind, but not in reality...the data rule out this [cause-and-effect] kind of universe....By virtue of their past, objects...can be part of a quantum-mechanically-entangled whole." A whole which shows evidence of having no disconnected 'parts'.

> *"Two things can be separated by an enormous amount*
> *of space and yet* not *have a fully* independent

existence. A 'quantum connection' can unite them, making the properties of each *contingent* on the properties of the other. Space does not distinguish such entangled objects. *Space cannot overcome their interconnection.* Space, *even a huge amount of space,* does not weaken their quantum-mechanical *interdependence.*"

"The quintessential quantum effect is entanglement," states physicist Vlatko Vedral, a University of Oxford professor, in a comprehensive article in *Scientific American* (June 2011). "Quantum behavior eludes visualization and common sense. It forces us to rethink how we look at the universe, and accept a new and unfamiliar picture of our world." Vedral continues:

> "General relativity assumes that objects have well-defined positions and never reside in more than one place at the same time—in direct contradiction with quantum physics. Many physicists, such as Stephen Hawking of the University of Cambridge, think that relativity theory must give way to a deeper theory in which space and time do not exist. Classical spacetime emerges out of quantum entanglements....

> "...space and time are two of the most fundamental classical concepts, but according to quantum mechanics they are secondary. The entanglements are primary. They interconnect quantum systems without reference to space and time....Gravity may not even exist, at the quantum level."

Anton Zeilinger and associates showed, in 1999, that molecules—not only atoms—exhibit wave-like properties. And in 2011, researchers (at Zeilinger's University of Vienna) observed quantum effects acting on a molecule of 430 atoms.

"Until the past decade, experimentalists had not confirmed that quantum behavior persists on a macroscopic scale. Today, however, they routinely do. These effects are more pervasive than anyone ever suspected. They may operate in the cells of our body."

In 2010, quantum effects were found to be involved in photosynthesis, in two species of marine algae.

"People have long wondered whether birds and other animals might have some built-in compass.... A bird's eye has a type of molecule in which two electrons form an entangled pair, with zero total spin....These molecules are indeed sensitive to magnetic fields, because of electron entanglement. According to calculations that my colleagues and I have done, quantum effects persist in a bird's eye for around 100 microseconds—which, in this context, is a long time.... Do any instances of larger and more persistent entanglement exist in nature? We do not know, but the question is exciting enough to stimulate an emerging discipline: quantum biology."

A mysterious intelligence is clearly at work, at the very least at the scale of molecules in particulate matter.

"You can set up the particles to have a total spin of zero even when you have not specified what their individual spins are. When you measure one of the particles, you will see it spinning clockwise or counterclockwise at *random*. It is as though the particle decides which way to spin, for itself. Nevertheless, no matter which direction you choose to measure the electrons, providing it is the same for both, they will always spin in opposite ways, one clockwise and the other counterclockwise.* How do they know to do so?"

Vedral concludes:

"...if a deeper theory ever supersedes quantum physics, it will show the world to be even more counterintuitive than anything we have seen so far."

Harvard-educated Lee Smolin is a professor of physics at Pennsylvania State University. In *The Life of the Cosmos*, he observes that it has been only a little more than a century since we've known "that the stars are organized into galaxies"; or, on a smaller scale, the nuclear structure of the atom. And science is also learning that

"...a philosophy which tells us to explain things by breaking them into parts will not help us when we confront...the things that have no parts."

Also,

"Why is it easier to conceive of a world structured by 'law' imposed from the outside" than to "imagine that the regularities of the world are all the result of processes of *self*-organization...without any need

---

* Bill Bryson: "It is as if, in the words of the science writer Lawrence Joseph, you had two identical pool balls, one in Ohio and the other in Fiji; and the instant you sent one spinning, the other would immediately spin in a contrary direction at precisely the same speed."

of an *external* intelligence...." Such new insights are relieving us of "myths that have been passed down to us from past generations."

Smolin says:

"There is an affinity between the ambition of theoretical physics and...of metaphysics. Both have often presumed that there is some absolute truth to be discovered about the world...the true essential—the true Being. Both...search for a transcendent and timeless actuality, beyond the *appearances* of the world.... A tradition that asserts that the world we see around us is not completely real....There is no way to...fly on the wings of *logic*, to ascend to the absolute world of what really is....In the end, perhaps, there must remain a place for mysticism."

Quantum mechanics is telling us, he says,

"...that the world can only be described...as an *entangled* whole....The real strangeness of quantum mechanics emerges when we apply it to systems that contain more than one thing....Whenever two systems [such as particles] have interacted, their description is tied together no matter how *far* apart they may be."

If two photons, flying away from each other, are measured (observed) at about forty-three feet apart (as in an early experiment), that space is 10^{11} power bigger than the atomic domain itself.

"The entangled nature of the quantum state reflects something *essential* in the world." That the properties of one particle are independent of another "is wrong—disproved by experiment. This makes it one of those rare cases in which an experiment can be interpreted as a test of a philosophical principle!...

> *"Given any one electron, its properties are entangled with those of every particle it has interacted with...* quite possibly from the very moment that particles were first created in our universe."*

Smolin says of the impact when he first learned this truth: he was "struck that there were atoms in my body that were entangled inextricably with atoms in the bodies of every person I had ever touched!"

If we were to learn more, at this point, about a particle—such as an electron—it would involve

> "...the relations between that electron and the rest of the universe....Any physical theory from this point on that represents progress *beyond* quantum mechanics must be an explicitly cosmological theory....If we want to give a complete description of an elementary particle, we must include...every particle it may have acted with....This means that we can only give a complete description of any *part* of the universe to the extent that we describe the *whole* universe.

> *"We who live in the universe, and aspire to understand it, are then inextricably part of the same entangled system. If we observe some part of the world, we become entangled with it in the same way that any two particles that interact become entangled; so that*

---

* "About 100 microseconds after the Big Bang, the protons and neutrons in your body were created." —Mark Whittle

a complete description of ourselves is impossible without incorporating the other."

Otherwise, we are,

"...then apparently in conflict with the results of experimental physics. Quantum mechanics works very well in *every* context in which it has so far been tested"; if we are to apply it on a universal scale, "there can be no place *outside* it for an [uninvolved] 'observer' to stand."*

We can also contemplate the possibility that both "time and change are illusions."

---

* Reminiscent of the Buddhist saying, "Nothing to stand on."

Consciousness

In *The Hidden Face of God,* Gerald Schroeder starts with the Big Bang, predecessor of time and space, when anything that could possibly have been "outside" and what was inside the "singularity" was merely one potential: "no divisions, no separations," he says.

"It formed our bodies *and* led to our thoughts," this "essence" (which means essential). He quotes physicist Freeman Dyson, "I do not make any clear distinction between mind and God."

"If indeed there is a universal consciousness," says Schroeder, "this could explain the interrelatedness of particles even when separated by large distances," as in quantum entanglement. As in the "double-slit" experiment, photons, electrons, even whole atoms passing through one slit "somehow know, and react accordingly, to conditions" or potential possibilities at the other slit; such as, "whether or not a conscious observer is present."

Someone living among and knowing only ice would not believe that he could heat his hands from steam created from that same ice. Equally hard for a person to comprehend is that when an electron, orbiting a nucleus, moves to the next higher or lower energy level, it is not on a gradient: it leaps from one orbit to another in zero time, without temporal transition. "The universe...does not comply with human reason," Schroeder says. "A kilogram of air has the same *gravitational* effect as a kilogram of steel."

"Physics has demonstrated," he states, "that a single substrate underlies all existence....The universe was born as an undifferentiated unity," originally perhaps much smaller than a mustard seed. "We are stardust come alive; and somehow, *conscious* of being alive....

"The beginning of our universe marks the beginning of time, space and matter....Whatever brought the universe into existence must predate time....In other words, it is eternal."

The four elements that produced the stars (suns) created the cauldrons across the skies which spewed out the remaining eighty-eight stable elements which form matter, some matter changing from one form to an entirely different form in the process (gold to lead, for example). Oxygen and hydrogen combined to produce something different from either gas: water. Some forms of this amazing product would act differently than the product itself: ice floats in water; floating ice serves as a thermal insulator for the water below, in effect preventing the oceans from becoming solid blocks of ice.

With as many as twenty-three atoms combining in molecules that make amino acids, a basis for living matter was formed. Attempts by scientists to combine similar molecular structures "has been one long study in failure."

A protein is a string of several hundred amino acids. They have something akin to an instinct for assembly. "But where did they get their smarts? Since when do carbon, nitrogen, oxygen, hydrogen, sulfur, phosphorus...have ideas of their own?"

In your head, there are a million billion dendrites, branching parts (the protoplasmic filaments) of a nerve cell. "I urge you to count to a billion, a million times....At one number each second, with no breaks for resting, that task will occupy you for the next 30,000,000 years."

"The truth," says Schroeder, is

"...that our *material* existence is more fiction than fact....Physics has touched the metaphysical realm within which our physical illusion of reality is embedded....Science has discovered a reality it had previously relegated strictly to the mystical. It has discovered the presence of the spiritual...."

He quotes Groucho Marx (caught in a delicate situation): "Are you going to believe me or your lying eyes?!"

"We are made of star stuff," said astronomer Carl Sagan. Both the human body and its world are composed of solids, liquids and gases provided by the cosmic Big Bang.

"There is nothing new to be discovered in physics now." British physicist and mathematician Lord (William) Kelvin is quoted as stating in 1894, notes *Quantum Enigma* by physics professors* Bruce Rosenblum and Fred Kuttner. Not a century had passed before the discovery that (as said by German theoretical and nuclear physicist Werner Heisenberg) "atoms, or elementary particles, themselves are not real: they form a world of potentialities or *possibilities*, rather than one of *things* or *facts*."

The authors:

---

* University of California at Santa Cruz.

"What does it say about things *made* of atoms?...Is it that quantum theory does not apply to big things? No....The workings of *everything* is quantum mechanical."

This includes a quantum effect known as "superposition," in which a quantum system can be said to exist in two states at once, such as potentially being in two locations.

"Recall our atom hitting a partially-reflecting/ partially-transmitting mirror, and ending up with *half* its waviness captured equally in each of two *separate* boxes. (According to quantum theory, the atom does not *exist* in one particular box *before* you find a *whole* atom to be in *one* of the boxes.) The atom is in a superposition state, simultaneously in *both* boxes. Upon *your looking* into one box, the superposition state (waviness) *collapses* into one *single* box. You will *randomly* find *either* a *whole* atom in that *one* box, *or* that box will be empty. (You can't choose which!) If you find the one box empty, the atom will be found in the *other* box."

Continuing:

"Increasingly large objects are being put into superposition states, put into *two* places at the same time. Austrian physicist Anton Zeilinger has done this with large *molecules* containing *seventy* carbon atoms,"

...and he anticipates doing so with mid-sized proteins.

"Truly macroscopic superpositions containing many *billions* of electrons have been demonstrated, where *each* is *simultaneously* moving in *two* directions."

Most startling of all quantum effects is what is known as the "delayed-choice" experiment, of particular interest to renowned physicist John Wheeler. He proposed a set-up where the *choice*, of whether to observe for outcome A or instead outcome B, was not made until after the particle experiment was underway: for example, if plan A was (subsequently) chosen, the particle would need to have presented as, say, "spin up"; and if plan B was chosen, "spin down." What if the experimenter didn't choose which property to observe, until after a particle had passed a certain point of "phase" no-return; did the particle determine beforehand which experiment would be enacted?

> "Wheeler's experiment was done with photons.... Quantum theory's prediction, that the *later* choice of experiment determined what the photon did *earlier*...was confirmed....Quantum theory is saying that our later *choice* of observation *creates* the particle's earlier history—we 'cause' something *backward in time!*"

The authors state:

> "There is a universal connectedness....Any 'objects' that have *ever* interacted *continue* to *instantaneously* influence *each other*....Mystics have talked of 'reality' and 'separability'—or its opposite, 'universal connectedness'—for millennia."

When an experiment proves universal connectedness between twin-state photons, it

> "...means a lack of reality or separability for *everything* such photons could possibly interact with. That is *everything*....

"The experiments showed that the properties of objects in our world have an *observation-created reality*; or that there exists a universal connectedness; or both."

"...[if] a better theory might supersede quantum theory [it] must *also* describe a world without separability....The experiments show that... *connectedness* can extend, beyond the photon pair, to *macroscopic* things....In principle, *any* two objects that have *ever* interacted are *forever* entangled... even if the interaction is through each of the objects having interacted with a *third* object. In principle, our world has a universal interconnectedness.... Quantum theory has no *boundary* between the microscopic and the macroscopic....

"In this most general sense, one can argue that the findings of physics *do* support the thinking of ancient sages. (When Bohr was knighted, he put the Yin-Yang symbol in his coat of arms.) [Quantum's] strangeness has implications beyond what we generally consider physics."

John Wheeler: "There is a strange sense in which this is a 'participatory universe.'"

England's Astronomer Royal, Martin Rees: "It does not matter that the observers turned up several billion years later. The universe exists because we are aware of it."

Physicist Freeman Dyson:

"It is conceivable...that life may have a larger role to play than we have imagined....The design of the *inanimate* universe [of matter and energy] may not be as detached from the potentialities of life and intelligence as scientists of the twentieth century have tended to suppose."

Cognitive scientist Donald Hoffman: "I believe that consciousness and its contents are *all* that exists."

The authors of *Quantum Enigma* point out that our interest in consciousness leads us to biology; and that leads us to chemistry; which leads us to "the interactions of atoms obeying quantum physics." And physics rests on what we know of the universe. What we know of it right now is that consciousness can have an effect on atoms. "In some basic sense, physics rests on the phenomenon of wave-function collapse by conscious observation." They also say, "If there's a *mind* that's other than the physical *brain*, how does it communicate with the brain? This mystery recalls the connection of two quantum-entangled objects." Physicist Francis Crick is quoted: "Your sense of personality and free will are, in fact, no more than the behavior of a vast assembly of nerve cells and their associated molecules." The authors add: "If so, our feeling that consciousness and free will are something beyond the mere functioning of electrons and molecules is an illusion." The formation of molecules of hydrogen and oxygen produce a surprising phenomenon that we know as "wetness," as in water: perhaps, they suggest, this gives us a hint of another peculiar phenomenon, "consciousness."

If the brain was to receive "influence" (a term of Bohr's) from the quantum field (or vacuum, as it's sometimes called), what would that say about free will? Recent research on decision making demonstrates that, at the neuronal level, we often make a commitment to a course of action even before we become *consciously* aware of it.

"In the early 1980s, Benjamin Libet had his subjects flex their wrist at a time of their free choice, but without forethought. He determined the order of three critical times: the time of the 'readiness potential,' a voltage that can be detected with electrodes on the scalp almost a second before any voluntary action actually occurs; the time of the wrist flexing, and the time the subjects reported that they had made their *decision* to flex (by watching a fast-moving clock).

"One might expect the order to be (1) decision, (2) readiness potential, (3) action. In fact, the readiness potential *preceded* the reported decision time. Does this show that some deterministic function in the brain brought about the supposedly free decision? Some, not necessarily Libet, do argue this way."

Through both consciousness and quantum entanglement, is it possible that there is some connection with the *creation* of the universe?

"Quantum theory has 'observation' creating the properties of microscopic objects. Physicists generally accept that, in principle, quantum theory applies universally. If so, all reality is created by our observation. Going *all* the way, the 'strong anthropic principle' [a theory] asserts the universe is hospitable to us because *we* could not create a universe in which *we* could not exist....Quantum cosmologist John Wheeler, back in the 1970s,... asked 'Does looking back [to the Big Bang] "now" *give* reality to what happened "then"?'"

How likely is it that the universe appeared by mere chance, irrespective of us "observers" who happen to flourish in it? "To produce a universe resembling one in which *we* can live, the Big Bang had to be finely tuned. How finely? Theories vary. According to one, if the initial conditions of the universe were chosen randomly, there would only be one chance in 10^{120} (that's one with 120 zeros after it) that the universe would be livable. Cosmologist Roger Penrose has it vastly more unlikely. The exponent he suggests is 10^{123}. By any such estimate, the chance that a livable universe like ours would be created is far less than the chance of randomly picking a *particular* single atom out of all the atoms in the universe!*

"Can you accept odds like that as a coincidence? It would seem more likely that something in yet-unknown physics determines that the universe *had* to start the way it did."

Freeman Dyson: "Life may have succeeded—against all odds—in molding the universe to its purposes."

"Scientists are now finding that there are ways in which the effects of microscopic entanglements 'scale up' into our *macroscopic* world....Some

---

* "There are something like ten million million million million million million million million million million million million million million (1 with eighty zeroes after it) particles in the region of the universe that we can observe."
—Stephen Hawking

scientists suggest that the remarkable degree of coherence displayed in *living* systems might depend (in some fundamental way) on quantum effects, like entanglement,"

so states Dean Radin in *Entangled Minds*.

"Others suggest that conscious awareness is *caused*, or related in some important way, to entangled particles in the brain. Some even propose that the entire universe is a single, self-entangled object. The idea of the universe as an interconnected whole [has] been one of the core assumptions underlying Eastern philosophies....Western science is slowly beginning to realize that *some* elements of that ancient lore might have been correct."

As Radin says,

"The bottom line is that physical reality is connected in ways we're just beginning to understand....Today we know that entanglement is not just an abstract *concept*....It has been repeatedly demonstrated as fact, in physics labs around the world since 1972.... Entangled connections are proving to be more pervasive and robust than anyone had previously imagined."

As stated in *New Scientist* magazine in March 2004:

"Physicists now believe that entanglement between particles exists everywhere, all the time; and have recently found shocking evidence that it affects the wider, 'macroscopic' world that we inhabit."

Organic molecules with an array of atoms have been successfully entangled; as have clusters of six entangled photons; it has been determined among the atoms of chunks of salt about ⅜ of an inch square that were entangled,

> "...photons, shot through sheets of metal, have been shown to *remain* entangled after punching through to the other side. Photons also remain entangled after being sent through 50 kilometers of optical fiber: and while being transmitted through the open *atmosphere*."

In the second edition of *Quantum Enigma,* Rosenblum and Kuttner report on other quantum developments that have come to light, such as these two accounts:

> "In 1997, researchers at MIT put a clump of *several million* sodium atoms, at low temperature, in a quantum state called a Bose-Einstein condensate. They then put this single clump two places at once, separated by a distance larger than a human hair. That's a small separation, but it's a macrosscopically seeable one. The *whole clump* was in both places at once."

> "A March 2010 article in *Nature News* is titled 'Scientists Supersize Quantum Mechanics: Largest Ever Object Put into Quantum State.' The object was a metal paddle only a thousandth of a millimeter long, but visible to the naked eye in the same way you can see a tiny dust mote in a sunbeam. The little cantilever was cooled to an extremely low temperature until it reached the most motionless state permitted by quantum mechanics, essentially standing still. It was then 'excited' to be in a superposition of that motionless state and *simultaneously* in a vibrating state. The paddle was moving and not moving at the same time."

Dean Radin points out,

"...there's no theoretical limit to how large an entangled object can be....*We're* still thoroughly permeated by entangled particles. Physicists have even speculated that entanglement extends to everything in the universe....Some, further speculate that empty space, the quantum vacuum itself, may be filled with entangled particles.* Such proposals suggest that despite everyday appearances, we might be living within a holistic, deeply interconnected reality. To be clear, these speculations are being proposed by traditional physicists, not by starry-eyed New Agers or mystics."

He declares,

"...we now know that fundamental properties of the world are not determined *before* they are observed....The common-sense assumption that ordinary *objects* are entirely and absolutely separate is *incorrect*....Unmediated 'action at a distance,' in quantum reality, is *required*.

"The new reality has dissolved *causality*, because the theory of relativity revealed that the fixed arrow of *time* is an *illusion*, a misapprehension sustained by the classical assumptions of an absolute space and time....The new reality has abandoned the assumption of *continuity*, because the fabric of quantum reality is discontinuous; at small scales, space and time are neither smooth nor contiguous. And finally, absolute *determinism* has been fatally challenged, because it relies on the assumptions of

* Astronomy professor Mark Whittle: "Particles are present, in a sense, in latent form in the vacuum of space."

causality, reality, and certainty; none of which exist in absolute terms anymore."

Continuing:

"Quantum reality is holistic; and, as such, any attempt to study its individual 'pieces' will give an incomplete picture....Few physicists today doubt that quantum theory[*] provides an accurate description of the observable world...[it] is so preposterously precise...the either/or logic of common sense no longer holds, in the quantum world. Until our language—and logic—evolve to (more easily-grasped) complimentary [interrelated] ideas, it's likely that we'll continue to experience confusion and paradoxes [apparent self-contradictions]."

One view among scientists today is

"...that consciousness is the fundamental ground state—more primary than matter or energy, [which] resembles ideas originating from Eastern philosophy and mystical lore. But a notable subset of prominent physicists, including Nobel laureate physicists Eugene Wigner and Brian Josephson, John Wheeler, and John von Neumann have embraced concepts that are at least mildly sympathetic to this view. Physicist Amit Goswami, from the University of Oregon, has strongly promoted this view."

The results of Bell's Theorem have "been described as the most profound discovery in science"; as late as 2004, the experiment was repeated yet again; over an effect distance of thirty miles. Physicists Abner Shimony (a professor of

---

* Physicist Anton Zeilinger: "Predictive power [of quantum mechanics] is unmatched by *any* other scientific theory."

both physics and philosophy) and John Clauser, involved in the earlier research, are quoted: "The conclusions from Bell's Theorem are philosophically *startling*...."

Radin notes, "in principle, any physical object" could be used in the experiment: "billiard balls, or humans." When finally you understand what this experiment tells us,

> "Your gut suddenly drops....[The word] *profound* isn't strong enough....Physicists are either thrilled or disturbed (sometimes both)*....The experimental evidence has now *convinced* the majority of physicists...something unaccounted for is *connecting* otherwise 'isolated' objects."

He quotes one of the founders of quantum theory, Austrian physicist Erwin Schrödinger, who coined the term entanglement:

> "Hence this life of yours which you are living is not merely a piece of the entire existence, but is, in a certain sense, the *whole*; only this whole is not so constituted that it can be surveyed in one single glance." (Or "observation"?)

Radin concludes:

> "It would be astonishingly unlikely to find that one small domain, the one that our bodies and minds happen to inhabit, are somehow *not* best described as quantum objects. As historian of science Robert Nadeau and physicist Menas Kafatos (both from George Mason University), describe in their book *The Nonlocal Universe*: 'All particles in the history of the cosmos have interacted with other particles,

---

* "We should trust what these [quantum] theories have to say." —Professor Mark Whittle

in the manner revealed by the Aspect experiments. Virtually everything in our immediate physical environment is made up of quanta that have been interacting with other quanta in this manner, from the Big Bang to the present....Also consider... that quantum entanglement grows exponentially with the number of particles involved in the original quantum state, and that there is no theoretical limit on the number of these entangled particles.'"

Radin:

"We don't know how big an 'influence' has to be, to cascade our brain states into one set of subjective experiences, versus another." But the possibility has been suggested that "human experience is indeed a part of the quantum reality."*

While our comprehension has only been recent, what we're comprehending is an inseparable reality that is far more ancient than even man's earliest intuition of Oneness—and may be the producer of that very intuition.

Physics concerns itself primarily with defining relative reality through finite measurements (or "computations"; for example, supplying hard numbers in the equation $E=mc^2$). "Classical" physics operates at the macroscopic level (for example, matter larger than a single molecule); "quantum" physics at the microscopic level, mainly atomic and sub-atomic scales.

It is a recognized fact in science today that if a scientist is "measuring" (subatomic) photons—for instance—as

---

* "The external world and consciousness are one and the same thing." —Erwin Schrödinger

particles, they will *present* as particles; if measuring for *waves*, they will appear in wave form. Our bodies, including our brain, are composed of subatomic particles.

As a consequence of the experimental proof of Bell's Theorem (also referred to as Bell's inequalities; and sometimes as EPR, after Einstein-Podolsky-Rosen who were catalysts for Bell), it is now scientifically recognized that quantum particles/waves, once (observed to be) "entangled", can *affect* one another—though remote from one another in space—*simultaneously*. This recent revelation suggests a supernatural intelligence which must be present *universally* ("non-local," as the physicists say).

Is the uncanny intelligence, which is displayed by subatomic particles, an inherent element in all that is saturated by them—which is to say, the entire *cosmos*? Is our own individual consciousness thus inseparable from the intelligence *governing* the cosmos, through every entangled (interconnected) quanta?

Physicist Roger Penrose,[*] a colleague of Stephen Hawking, wrote *Shadows of the Mind* to initiate a tentative exploration of the reaches of quantum *interconnectedness*. He says there are parts of the neurons in our brains that are configured in such a way that they are likely engaged at the quantum level.

As a science article has put it, Penrose proposes that consciousness "is a byproduct of quantum processes operating in the brain...[that] can allow quantum phenomena to *influence* how neurons behave."

The three pounds of soft tissue that is the human brain amounts to about two percent of the body's mass, yet utilizes twenty percent of the energy supplied by the body's

----

[*] Penrose is a mathematical physicist, at Oxford University.

volume of blood. The brain functions with just fifteen watts of power—about as much as a refrigerator light bulb.

Comprising the brain are more than a hundred billion neurons (that's 10^{11} power); during the forty weeks of gestation, approximately five thousand neurons are created in the fetus every second, but of the resultant 150 billion, some atrophy.

Of the (ten thousand trillion) cells in the human body, the neurons are capable of communicating in ways which result in more than a hundred trillion interconnections in the cerebral cortex of the brain, the center of cognition. There are about a billion such connective synapses per cubic centimeter in the frontal cortex.

A neuron is, says science writer Susan Kruglinski, "an elaborate processor, powered by neurotransmitters." Each of these cells "can receive up to 150,000 contacts from other neurons" via more than fifty varieties of neurotransmitters, that may be arranged in packets of five thousand molecules each.

These electrically-charged chemical transmitters can be activated by as little as 0.1 volt—1/100,000 the strength of a static shock from a carpet—and can generate "action potentials" in a neuron at the rate of three hundred per second. These electric charges travel via a network of connective axons, at up to 270 miles per hour. In a receptor cell, they can create a change that is measured at a hundred million ions per second.

What interested Penrose, in this context, is that running along the length of the axons are microtubules that transport neurotransmitter molecules. But otherwise,

> "The tubes themselves appear to be empty—a curious, and possibly significant, fact in itself.... 'Empty,' here, means that they essentially contain

just water.... Not at all like ordinary water, with molecules moving about in an incoherent, random way. Some of it...exists in an *ordered* state...."

Think of a crystal.

These microtubules have a diameter of about 25-30 nanometers (nm: one billionth of a meter), compared to their relatively long length of a millimeter (mm: one thousandth of a meter; .03937 inch) or more.

"It seems not unreasonable to suppose," Penrose says, that the microscopic size[*], and other conditions of the conduits, "would strongly favor the possibility of quantum-coherence oscillations within...the tubes."

"Quantum coherence" describes large numbers of subatomic particles collectively cooperating as a single quantum state. Quantum behavior can include electrons, photons, atoms and molecules.[†] Coherence can occur in the action of a laser, and also in what is known as a Bose-Einstein condensate. The "wave function," in such case, is "of the kind that would be appropriate for a single particle (Penrose)," but the collective behavior of all particles is as a whole—and this at a macroscopic (visible) level. Penrose says there is "a *distinct* possibility of quantum coherence having a genuinely significant role to play in biological systems."

He had, in 1987, already described light-sensitive cells in the retina of the eye (which technically is an extension of the brain) that have the capability of responding to even as few as one isolated photon. This led him, at the time, "to speculate that there might be neurons in the brain, proper, that are also essentially quantum-detection devices."

---

[*] If an atom was the size of a marble, says Mark Whittle, our brain would, proportionally, be twice the size of the earth, and synapses would be thousands of miles long.

[†] More than 1,400 types of molecules are involved in synaptic transfers.

In 1992, renowned neurophysiologist John Eccles posited quantum effects in synaptic actions, saying that an appropriate location would be what is called the presynaptic-vesicular grid, described technically as "a paracrystalline hexagonal lattice in the pyramidal cells of the brain."

What is significant to Penrose is that the interaction of subatomic particles known as "entanglement" is "an effect that does not fall off with distance," as does gravitational or electric attraction; it is oblivious of space separation and time transition. The effect holds "no matter how distant from each other" the interconnected particles may be, and it appears that such condition persists forever. And an implication is that everything in the universe is similarly affected. "So long as these entanglements persist," says Penrose, "one cannot, strictly speaking, consider any object in the universe as something that is on its own." Phenomena such as this "we must take seriously, as true aspects of the behavior of the constituents of the world in which we live."

His conclusion:

> "Let us then accept the possibility that the totality of microtubules in the...large family of neurons in our brains may well take part in 'global' quantum coherence...across the brain."

Summation

The mystic	The physicist
"Suchness is neither that which is existence, nor that which is non-existence; nor that which is at once existence and non-existence; nor that which is not at once existence and non-existence."	"If we ask, for instance, whether the position of the electron remains the same, we must say 'no'; if we ask whether the electron's position changes with time, we must say 'no'; if we ask whether the electron is at rest, we must say 'no'; if we ask whether it is in motion, we must say 'no'."
Ashvaghosha	Robert Oppenheimer

A ground-breaking book, for science as well as the general public, was Fritjof Capra's *The Tao of Physics*. Capra took nearly a year and a half out of his career as a physicist to write this book. Major publishers in New York and London were among dozens to reject it, but it soon became an international bestseller.

To begin with, he cites the Hindu concept that God becomes the *world* which, in the end, becomes God again (which could be said of the *self*, as well). Somewhere in this process, self-conceived "individuals" have the opportunity, chance, to recognize themselves for who they truly are: "In this state, the false notions of a separate self have forever disappeared,* and the oneness of all life has become a constant....The intellect is seen merely as a means to clear the way for a direct mystical experience...where reality appears as undivided and undifferentiated...."[†]

The essential nature of reality is emptiness—

"...not a state of mere nothingness, but is the very source of all life and the essence of all forms.... It *pervades* all material things in the universe....

"...the process of enlightenment consists merely in becoming what we already are from the beginning.... [It is] the experience of all phenomena in the world as manifestations of a basic oneness... of the same *ultimate* reality—also one of the most important revelations of modern physics."

What the sages knew through the intuitive instinct, physicists are affirming through the process of experiment.[‡] One starts from focusing on the immaterial, the other from focusing on the material. What the mystic discovers is intangible; what the physicist discovers—in the seemingly tangible—is likewise intangible. Both discoveries have to be described within the limits of inadequate language.

To speak of such fundamentals as space and time as absolute and independent entities is inadequate (as sages have maintained), because both concepts have now been

* "The true value of a human being," proposed Einstein, "is determined by the measure and the *sense* in which he has attained liberation from the self."
† Physicist Nick Herbert: "Reality is an undivided wholeness."
‡ Both the word *science* and the word *sage* are based on Latin roots: *sciens,* "to know," as in to discern; and *sapiens,* "to know," as in to taste.

abolished as physical principles; anything which we could call a "time" measurement would vary in different parts of our universe: as Buddhism has asserted, "*all* things change," everything is (at best) relative.

As Capra says, "The basic unity of the cosmos manifests itself not only in the world of the very small, but also in the world of the very large": "sameness" would be the Buddhist term.

"Being transient manifestations of the void [physics: "field"], the things in this world do not have any fundamental identity [of their own]." This, Capra points out, is also true of the "self." "Particles cannot be separated from the space surrounding them": rain drops, for example, do not condense out of a vacuum; space not only manifests the single raindrop but coordinates the activity of its counterparts in the function of precipitation. Capra quotes a science text:

"The field exists *always* and *everywhere*.... It is the carrier of all material phenomena [and their interactions]. It is the 'void' out of which the proton *creates* the pi-mesons."

The coming into being, and fading away again, of the particles "are merely forms of motion of the *field*," not activities initiated, or directed, by the particles themselves. That the particle is not in control of its circumstances, as *isolated* behavior, is demonstrated by the fact

"...that 'virtual particles' [a particular form] can come into being *spontaneously* out of the void, and vanish again into the void, without any nucleon (or other strongly interacting particle) being present.... The vacuum [or void] is far from empty..."

...it is dynamic; a field is defined as "a physical *quantity* in a spatial region." Capra says it contains the potentiality for forms, their formulation.*

The quantum field is viewed as the fundamental ground of being, present everywhere in space; the presence of particles are localized events, like ripples, in the field, "concentrations of energy, which come and go"; their reality as an entity is merely a temporary phenomenon.† He quotes Einstein: "the field is the only reality"; it does not come and go.

Capra says of both Eastern sage and quantum physicist, the view is of "physical things and phenomena as transient manifestations of an underlying fundamental" reality, "the only reality," called a "field" by one and a "ground of being" by the other. However, while a field might be regarded as a form or entity, the ground of being is formless, it being from which forms arise—"the *essence* of all forms," the essential nothingness that gives somethingness its identity (the two together being called, in Buddhism, "suchness"). In that nothingness is a potential for everything.

And, from which the definable forms arise, like particles, to that they return. The vital process entails presence, interaction, transmutation. *Being* and *change* are aspects of one process.

---

* Astrophysicist Mark Whittle: "A quantum field extends across the universe, and permeates everything. Think of such a field as [a form of] knowledge embedded in space....Where does the knowledge reside to do all [cosmic] things in just the right way?...The knowledge is somehow embedded in the vacuum field....So, in some sense, the knowledge to do [quantum interactions] is located at *every* point in *space*."

† "Particles can come into being and be gone again in as little as 0.000000000000000000000001 second (10^{-24}). Even the most sluggish of unstable particles hang around for no more than 0.0000001 second (10^{-7})."— Bill Bryson

We tend to think of "life" as a kind of experience.* Matter on the atomic level, says Capra, is "always in a state of motion. Particles do not stand around passively." The closer we examine matter, even metal or stone, "the more *alive* it appears." Actively.† "Movement and rhythm [harmony] are essential properties of matter," the rhythm having to do with co-operative patterns, which are ever-changing from existence to nonexistence of form. The One manifests as many; the many dissolve into the One: Chuang-tzu called it "transformation and change."

"This ultimate essence," remarks Capra, "cannot be separated from its multiple manifestations"; movement and being.

"In atomic physics, we even have to go beyond the concepts of existence and nonexistence...which is most difficult to accept." It brings to mind paradoxical statements in such as the Upanishads: "It is within all this, and It is outside of all this."

When we speak of Oneness, we must speak of the unity of—actually, the transcendence of—conceived opposites. When we deny fundamental Oneness and assert, for example, the reality of "rightness," we establish a conflicting condition, "wrongness"; Buddhists call this "mutually-arising." Chuang-tzu: "When the 'that' and the 'this' cease to be opposites, that is the very essence of Tao." Opposites are interdependent for their very definition, or "meaning."

---

* Physicist Paul Davies: "The problem is that there is no real definition of life. Living systems are examples of organized matter and energy, at extreme levels of complexity, but no boundary exists between the living and nonliving. Crystals, for example, are highly-ordered structures which can reproduce themselves, yet we do not regard them as living. Stars are complex and elaborately-organized systems, but are not normally thought of as alive. It could be that we are too narrow-minded in our vision of life...."
† In the Gospel of Thomas, Jesus ("before Abraham, *I am*") is quoted, "Cleave a piece of wood, I am there; lift up the stone, and you will find me there."

153

When they are taken to be independent realities, conflict ensues. *Divisiveness* is the root of strife (and striving).

The difficult matter for spiritual aspirants to comprehend is that, as a sutra puts it, "form is emptiness, and emptiness is form": these are not mutually exclusive conditions but merely perceived ("named," as they say in Buddhism) aspects of a singular actuality. When recognized as such, "Oneness" is the present condition; transcendence of "name," or conceptual identity, is inextricably involved.

"A subatomic particle," says Capra, "is not an isolated object but rather an *occurrence*, or event." Eventually it is an interconnection to something related to its presence. "The *structure* of a hadron, therefore, is not understood as a definite arrangement of constituent *parts*, but is *given* by *all* sets of particles which may interact with one another to form the hadron under consideration."

In an application of physics, "All particles are seen as intermediate states, in a network" of interactions, "better described as an event...." Capra quotes Buddhist scholar D.T. Suzuki, "Buddhists have conceived an object as an event, and not as a thing or substance."

Most subatomic particles dis-integrate into other particles. A particle at any particular moment can only be described as having "a tendency to exist," with transformation "a tendency to occur." Ashvaghosha: "[With] all forms of material existence...we cannot describe any degree of (absolute or independent) reality to them."

Being interdependent, no part of the universe is more important than any other. To the extent of what is essential, every part bears the importance, or "meaning," of all others. As every reality is in the all, the all is in every reality. Thus, it has been said: "In every particle of dust, there are present Buddhas without number." And thus D. T. Suzuki can state of the enlightened mind, "The Buddha...lives in a spiritual world [or world of "spirit"] which has its own rules," as does the quantum reality. As Capra notes of physics, a "concept can be given a *precise* mathematical meaning but is almost impossible to *visualize*."

Such is the paradox in the nondual teachings, that all that is relative is within the Absolute, simultaneously as the Absolute is within all that is relative. To realize *this* is to realize that, in the ultimate reality, there is no division; not anything is apart from anything else (thus, "distance," and "time" to traverse it, have no meaning—or are "illusions"). Hence, Blake could say, "To see a world in a grain of sand. . . ." One who can accommodate that, can see the observer in the observed (or vice versa).

The potential interaction of a particle, Capra points out, can only be understood in terms of a relationship between its presence and the presence of its observer, interdependently. If we want to observe a particular particle, we may first have to bring it into being (such as in a particle collider). In doing so, we may create or destroy other particles. If we have a different intent, the condition of the observer and what is observed will both change interdependently. In the same way that what is observed is arbitrarily defined by us, so too is how we come to define the observer, our self. Different observers do so in different ways, so there is nothing non-elastic about what is defined as a self: it's relative, just as whether a quanton presents as a particle or wave depends upon the manner in which one decides to observe it. What you see is what you get.

Capra: "The structures and phenomena we observe in nature are nothing but *creations* of our measuring [comparative] and categorizing *mind*. That this is so is one of the basic tenets of Eastern philosophy." Change your state of consciousness, and what you get is what you see. To have an attachment to any perceived reality is to go astray. Buddhists call this "ignor-ance," ignoring a deeper truth. Ashvaghosha: "All phases of the defiled mind are thus developed."

Physicist John Wheeler said that instead of scientific observer, we should say "participator." He added, "In some strange sense, the universe is a participatory universe"; he was likely thinking of wave-function collapse.

To this, Capra adds: "*Mystical* knowledge can never be obtained just by observation, but only by full participation with one's whole being." The Upanishads say, "Where everything has become just one's own self, then whereby and whom would one see?" True unity means, says Capra, "one's 'individuality' *dissolves* into an undifferentiated oneness... and the notion of [separate] 'things' is left behind." This is what D.T. Suzuki calls the "Absolute point of view"; a point of view which an increasing number of scientists, like Capra, are coming to contemplate.

John Polkinghorne received a doctorate in theoretical physics from Cambridge University, and held a professorial chair there. He also performed research with Nobel laureate Murray Gell-Mann,* which discovered the quark as the kernel of the atom. At the end of a 25-year career in 1979, he was ordained as an Anglican priest.

At around his eightieth birthday, he engaged in a meeting of physicists, at Oxford University in 2010, who had an interest in (what *Discover* magazine called) the "physics of the divine." As an associate of Polkinghorne, physicist Bob Russell, explained, "science can be a spiritual experience.

* Gell-Mann gave his discovery, the quark, its name.

For some scientists, it's about reading the mind of God."
Physicists like Russell (the magazine said) "concluded
that the best place to seek scientific support for God is in
quantum mechanics."

The magazine also referred to quantum physicist Antoine
Suarez:

> "Most physicists accept entanglement as just one
> more counterintuitive reality of quantum physics.
> But Suarez claims entanglement tests conducted
> with real photons in the lab suggest that quantum
> effects must be caused by 'influences that originate
> from outside of space-time'....

> "Whatever causes the twin photons to behave in
> the same way, it must work independently of time.
> 'There is no story that can be told within the
> framework of space-time that can explain how these
> quantum correlations keep occurring.' Suarez says.

> "These results have intriguing philosophical
> implications, he notes, especially for the spiritually
> inclined. 'You could say the experiment shows that
> space-time does not contain all the intelligent entities
> acting in the world, because something outside of
> time is coordinating the photons' results.'"

The Dalai Lama, who sometimes conversed with David
Bohm, stated in *The Universe in a Single Atom*, concerning
quantum physics:

> "To a Mahayana Buddhist, exposed to Nagarjuna's
> thought, there is an unmistakable resonance
> between the notion of emptiness and the new

physics. If, on the quantum level, matter is revealed to be less solid and definable than it appears, then it seems to me that science is coming closer to the Buddhist contemplative insights of emptiness and interdependence. At a conference in New Delhi, I once heard Raja Ramanan, the physicist known to his colleagues as the Indian Sakharov, drawing parallels between Nagarjuna's philosophy of emptiness and quantum mechanics.

"After having talked to numerous scientist friends over the years, I have the conviction that the great discoveries in physics (going back as far as Copernicus) give rise to the insight that reality *is not as it appears* to us."

His conclusion:

"All things and events—whether material, mental, or even abstract concepts like time—are devoid of objective, independent existence."

Writing about scientific advances in the study of human consciousness, Jay Tolson (*U.S. News & World Report*, 10-23-06) declares,

"...new thinking in philosophy and theology is questioning the assumption of an absolute divide between mind and body, spirit and matter—an assumption that has long sustained many religious conceptions of the soul. Interestingly, these parallel developments in science and religion point to a new picture of reality—or maybe even recall older understandings implicit in traditions as ancient as Judaism or Buddhism—in which subject and object, mind and matter are more interfused than opposed."

Of cognitive theorist and philosopher Daniel Dennett, he writes:

"The big mistake, according to Dennett, is to think that there is some homunculus of a self sitting in the theater of the brain and observing, or even directing, the ongoing show."

The article continues:

"If this view is true, where is the self or identity on which even a broadminded religious believer might base his notions of the soul? Here Christians and others might turn to the wisdom of Buddhism, in which the self is correctly understood not as an entity or substance...."

Physicist Shimon Malin, in *Nature Loves to Hide*, basically explains nonduality.

"When I think about 'the One', I make a distinction between 'One' and 'not-One': I am thinking of *two* items...not in accord with the One itself. Similarly, even the distinction between 'being' and 'non-being' does not apply to the One; its transcendence is so complete that I cannot even say, 'The One is'."

He quotes physicist Erwin Schrödinger:

"Inconceivable as it seems to ordinary reason, you—and all other conscious beings as such—are *all in all*. Hence, this life of yours which you are living is not merely a *piece* of the entire existence, but is, in a certain sense, the *whole*; only, this whole is not so constituted that it can be surveyed in one single glance. This, as we know, is what the Brahmins express in the sacred, mystic formula which is yet really so simple and so clear: *Tat tvam asi*....[*That thou art.*]"

And Malin quotes Roman philosopher Plotinus (c. 250 A.D.): "There is no two...(man) is merged with the Supreme, sunken into it, one with it...which is to be known *only* as one with ourselves."

Malin sums up:

"The challenge of coming to the ineffable knowledge of *who I really am* is the same as the challenge of coming to the ineffable knowledge of *the One*. This is so because, ultimately, *I am the One*. If one accepts that 'the One' is a name indicating the real, nameless *source of what is*, rather than an abstract concept, then, oddly enough, the statement 'I am the One' can be proved: If I were not the One, then the level of 'the One' would have consisted of at least two items, me and the One, rather than there being truly one."

A contemporary of Einstein, English astrophysicist Sir Arthur Eddington:

"Is the ocean composed of water, or of waves, or both?...I think the ordinary, unprejudiced answer would be that it is composed of water. At least if we declare our belief that the nature of the ocean is aqueous, it is not likely that anyone will challenge us and assert that on the contrary its nature is undulatory.

"Similarly, I assert that the nature of all reality is spiritual, not material; nor a *dualism* of matter and spirit....

"Interpreting the term 'material' (or more strictly, physical), in the broadest sense, as that with which we can become acquainted through sensory experience of the external world, we recognize now that it corresponds to the waves, not to the water of the ocean of reality.

"My answer does not deny the existence of the physical world, any more than the answer that the ocean is made of water denies the existence of ocean waves; only, we do not get down to the *intrinsic* nature of things that way."

Austrian physicist Erwin Schrödinger:

"...there *is* only one thing; and what seems to be a plurality is merely a series of different aspects of this one thing—produced by a deception (the Indian *maya*). The same illusion is produced in a gallery of mirrors; and in the same way, Gaurisankar and Mount Everest turned out to be the same peak, seen from different valleys....The plurality that we perceive is only *an appearance*; *it is not real*. Vedantic philosophy...has sought to clarify it by a number of analogies, one of the most attractive being the many-faceted crystal, which—

while showing hundreds of little pictures, of what is in reality a single existent object—does not really multiply that object."

Fritjof Capra, excerpts:

"As always in Eastern mysticism, the intellect is seen merely as a means to clear the way for the direct mystical experience, which Buddhists call the 'awakening'. The essence of this experience is to *pass beyond the world of intellectual distinctions and opposites* to reach the world of *acintya*, the unthinkable, where reality appears as *undivided* and *undifferentiated* 'suchness'."

"The highest aim for their followers—whether they are Hindus, Buddhists or Taoists—is to become aware of the unity and mutual interrelation of all things; to transcend the notion of an isolated individual self; and to identify themselves with the ultimate reality. The emergence of this awareness—known as 'enlightenment'—is not only an intellectual act but is an experience which involves the whole person, and is religious in its ultimate nature."

"In this state, the false notions of a separate self have for ever disappeared and the oneness of all life has become a *constant* sensation. *Nirvana* is the equivalent of *moksha* in Hindu philosophy and, being a state of consciousness beyond all intellectual concepts, it defies further description. To reach *nirvana* is to attain awakening, or Buddhahood."

"The experience of oneness with the surrounding environment is the main characteristic of this (meditative) state. It is a state of consciousness where every form of *fragmentation has ceased*, fading away into undifferentiated unity."

"The *fragmented* view is further extended to society which is split into different nations, races, religious and political groups. The *belief* that all these fragments—in ourselves, in our environment and in our society—are really separate can be seen as the essential reason for the present series of social, ecological and cultural crises. It has alienated us from nature and from our fellow human beings."

In the final two chapters of *The Dancing Wu Li Masters*, writer Gary Zukav especially zeroes in on the essence of quantum mystery, and its implicit message:

"There is only one reality, and it is whole and unified. It is one....The phenomenon of enlightenment and the science of physics have much in common....

"'This' and 'that'...are different *forms* of the same thing. Everything is a manifestation...of 'that which is'....*We* are manifestations of that which Is. *Everything* is...[even] that which is *not* is that which Is. There is nothing which is not that... There is nothing *other* than that....In fact, we *are* that which *Is*....

"Bell's Theorem and the enlightened experience of unity are very compatible. [It] tells us that there is no such thing as 'separate parts'."

Zukov quotes Henry Stapp:[*]

"Bell's Theorem...shows that our ordinary ideas about the world are somehow profoundly *deficient,* even on the *macroscopic* level."

Zukov concludes: "Everything, even 'emptiness,' is that-which-is"; and both Being and Non-being: "There is nothing which is not that-which-is."

Steve Hagen is a scientist who was drawn to study Zen in order to better understand what was being revealed to him by physics. In *How the World Can be the Way it Is,* he quotes Zen master Shunryu Suzuki: "I have discovered that it is necessary, absolutely necessary, to believe in nothing... [that] which exists before all forms." Hagen states early on,

"There can be nothing outside absolute Oneness; it is boundless," without barriers of any kind. "Our ordinary mind only sees realities which are relative, and therefore fragmented....Modern science provides us with a very good example of boundlessness, however: the universe."

He continues:

"To be boundless means not to see something 'over there', as if it were apart from yourself...as if there were some locality completely separate from 'here.'"

And:

"The Absolute aspect...[is] often taken to be imaginary....There's nothing to *compare* it to, which accounts for why our...mind habitually

* A physicist at the University of California, Berkeley.

misses this aspect of Reality. When seeing the world as a collection of *parts*...we imagine boundaries *dividing* these 'parts'...such as making the distinction between *you* and *me*....Through our... dividing the world into *this* and *that*, we make it... less full of real meaning....We insist that the world must be *this way* or *that way*....And we wonder why it doesn't make sense...."

"An electron's position," here *or* there, "is not something which really exists, *until* we look for it." The so-named uncertainty* principle,

"...is an essential ingredient of physical reality....

"Without the consciousness of an observer, the stuff underlying this physical reality does not seem to *exist*....Things are instead weirdly blended with, or take their identity from, what they are not."

Hagen points out,

"If we take two subatomic particles (say, protons) and smash them together at extremely high speeds, we find that the two original colliding particles fly apart—along with two new *additional* particles [which] didn't exist anywhere, in time or space, *before* the collision."

Imagine smashing two watches together, and finding two more similar watches among the debris. "How substantial is *matter*—the book you're reading now, or the hand which holds it?"

---

* Astronomer Hugh Ross gives an example of uncertainty of relative position: "If we were to measure the position of an electron to within a millionth of an inch, we could determine that electron's position one second later to an accuracy no more precise than 1,500 miles!"

Further:

"The total amount of energy in the universe [the 'positive' energy in matter and the 'negative' energy of gravity] adds up to about zero. If we could put all of the universe in one place [e.g., on a scale], *it* would add up to zero too."

He quotes physicist Nick Herbert regarding the scientists' dilemma, "If 'quantumstuff' is all there is and you don't understand quantumstuff, your ignorance is complete." Herbert says the "alternation of identities" [such as that of particles] "is the major cause of the reality crisis in physics."

Hagen states: "We don't approach life from a perspective of Totality and Wholeness [but] by seeing *myself over here* and *everything else over there.*" Such divisive "reductionism" presents us with fragmentary views of existence: there is me, my brain, my mind, my thoughts, my actions, as if these were separate components.

There is, prior to these distinctions, simply consciousness. "Consciousness is necessarily antecedent to matter, as far as any experience of matter can provide." We were conscious before we ever conceived of "brain," "thoughts," "self," etc. Yet we continue to suppose "that matter precedes consciousness"; and we take objects to be of primal importance, though it is consciousness which creates objects as "entities"—things that have defined, individual existence, or "reality," and relationships between them (whether conceived as "material" or "immaterial").

Thus, the absolute Totality is reduced to "parts," in our cognition—such as *me* and *you.* This is the root of self-centeredness, self*ish*ness; conflict. It is also the root of *man* being estranged from *God.*

"There is no such duality. That there are *two*, and yet that there are *not two*, occur at once and in the same location,"* the relative appearing within the Absolute. This applies, as well, to "me" and "consciousness"; there is no individual "me" outside of consciousness. When the me (subject) speaks of "my consciousness" (object), duality is inevitably conceived. This self-division Hagen refers to as the "big bang" of separative origin.

All objects are quantum objects, but they do not appear to our mind's eye as extra-ordinary, similar to the way that you look in the mirror and do not recognize the product of inter-stellar processes. While the extra-ordinary and what we view as ordinary are the same, we overlook this reality. The macroscopic and the microscopic are not in two different universes.

To "measure," which is what scientists do, is to fragment reality. We want to become "conscious" of "reality," while consciousness has never been separate from reality. Without comprehending the nature of Wholeness, we aspire to discover Wholeness. We're still in shock to discern that a proton is both a particle and a wave, distinguishable only in accord with what we happen to be looking for.

"Bell's Theorem has led us to the discovery that... though we conceive of a *here* and a *there*, such conception is not supported...by experimental results...'two' which are *not* two [even in] the very fabric of time and space itself."

Hagen quotes Nick Herbert: "Bell's Theorem tells us that it is...a *reality* to be reckoned with." The irony is that, in his initial intention,

* Physics professor Richard Wolfson: "Two simultaneous events are the *same* event."

167

"Bell* aimed to *validate* the common notion that *I* am separate from *you* [but] ended up proving precisely the opposite...much to his surprise and chagrin!"

There is, Hagen says, just *this*—not *that*: "We misapprehend what we actually take in." We are very familiar with the relative; it's time we became acquainted with the Absolute which permeates it. It isn't the Absolute which is an "abstraction," it is the relative. The Absolute is present in the mirror we call consciousness, as it were, and the reflection (without lasting substance) is the relative images, the perceived "this" and "that."

"We're not seeing things as they are," Hagen says. "We're missing something"—what in the nondual teachings is referred to as the transcendence of *this* and *that* (the dualistic myopia).

For those who have transcended the dualities, "Actions that spring from an awareness of the Whole...is utterly beyond any everyday sense of...*right* and *wrong*, or pleasant and unpleasant....

"We will either act out of our confusion" [that conceptual distinctions are relevant], or respond to the Totality,

"...as it actually is, not as we would hope, desire, imagine, or conceive it to be.... To live and act from out of the whole, and not the part.... Then no prescription or set of commandments is necessary... because we have a clear view of Reality—and of the universe."

----

* The late John Bell was a particle theorist who mused, regarding physics (emphasis mine): "Suppose...we find an unmovable finger obstinately pointing *outside* the subject, to the mind of the observer, to the Hindu scriptures, to God, or even only gravitation? Would that not be very, very interesting?"

Zen master Shunryu Suzuki:

> The true purpose of Zen...
> is to see things as they are...
> and to let everything go
> as it goes.

"The basic *oneness* of the universe, as revealed by quantum mechanics, is also the central characteristic of the *mystical* experience," remarks Darling.* "Twentieth-century physics has finally caught up with the philosophy of the Far East." Niels Bohr made references to Buddha and Lao-tzu, and Werner Heisenberg to the wisdom of the Far East.

And, yet, in much of the orthodox scientific community, researchers

> "...proceed as if there were an objective world out there....At the heart of our traditional Western outlook is *dualism*....So, we tacitly assume that through our (mental) will we move our (material) bodies....[We] think in a dualistic way...[in] a world of apparently irreconcilable differences.

> "And one of our *principal* misunderstandings stems from the use of our words 'you' and 'I'....Our language forces us to...break down our experience of the world into composite elements....distancing the perceiver from the perceived....

* Darling, *Zen Physics.*

"At the subatomic level, all divisions and boundaries imposed by us on the universe are in fact illusory—including the split between mind and matter."

What we think of as facts—measures and numbers—are primarily abstractions. The aim of nondual realization is to integrate the observer and the observed in a direct and immediate way, bypassing all abstractions and conceptualization.

"The deep, latent message of quantum mechanics... is that there is a reality, beyond our senses, which eludes verbal comprehension or logical analysis."

"The wave and particle natures of light and matter are not mutually *exclusive*, they are mutually *inclusive*...aspects of reality....And so, one of the central principles of modern physics is coincident with...one of the most basic doctrines of the Eastern worldview...."

"Zen and physics, then, seemingly so different, are not so different after all," like waves and particles. "Quantum mechanics presently appears to be as profoundly paradoxical and enigmatic as Zen." A physics question can sound like a Zen koan: if a tree falls in the forest, when there are no ears to hear, does it make a *sound*?

"The 'sages' of both East and West now tell us [there] is the *one* true reality."

Darling goes on to surmise that "the very act of seeking may block or hinder the experiencing of enlightenment." Intent can have an effect on what manifests. Even if we say, "Stop seeking and start experiencing," that is a dualistic, either/or formulation. If we transcend the reductionist labeling of a defined state of seeking and a defined state called enlightenment, there is only a singular, undifferentiated condition which is a present actuality, a perception of "Oneness." Darling avers, "There is only one reality." That reality "is all about direct experience, unadulterated being....We are urged to lose ourselves and merge with the whole." He quotes Meister Eckhart: the omnipresence that we call Being is all things.

Darling notes that with subjects who've had near-death experiences, "in some cases, profound transcendent experiences apparently took place after the person had been pronounced clinically dead," in apparent circumstance of "the removal of the brain's restricting influence [when] the psychological walls of the self are broken down."

In summation, Darling states,

> "We shall never, in a billion years, be able to explain how the *brain* [material] gives rise to consciousness [immaterial]....Consciousness can never be *divorced* from matter [and thus] has to be seen in a radically new light....It is ubiquitous....a permanent, inherent property of the universe."

Because each of us experiences consciousness, we conclude that it is individuated. If moonlight shines atop twenty buckets of water, would we conclude that moonlight is a separate property of each container? No:

> "...reality is an unbroken unity, and *within* this unity are aspects of the whole [so-called "individuals"] that *think* of themselves as being 'separate'."

Because we have a sense of being, we associate that with being one "self"—"within the undivided totality of what is real...an overall system that actually has *no parts.*"

> "Human beings...have been created by the universe....And now they are beginning to see beyond the 'self,' to the truth of their condition.... The signs of emergence of a...cosmic perspective are evident [such as] in the esoteric philosophy of quantum physics.... The only reality that exists, it is becoming clear, is right in front of us; nothing is hidden.... We have peered inside ourselves...in search of a soul and have found nothing.

> "The universe is one, and to see it as such is the goal...of science."

Our bodies, our brains, are

> "...composed of atoms whose nuclei were manufactured inside the intensely hot cores of giant stars that exploded in the remote past....We are nothing less than the universe in dialogue with itself."

Darling quotes "a hardened pragmatist," the late biologist J. B. S. Haldane:

> "It seems to me immensely unlikely that mind is a mere byproduct of matter....I am already identifying my mind with an absolute, or unconditioned mind... and the more I do so, the less I am interested in my private affairs...."

Darling suggests, "at death, we effectively rejoin the unbroken sea of consciousness." The brain's function is the limitation of things to objective status, such as "my"

172

consciousness, dualistically pursuing subject and object, so that the organism can operate in a relative world of bodily needs and environmental supplies. But "selves come and go, as brains come and go....We don't really own or exert will over our bodies and minds; we are simply aspects of an endlessly-unfolding process....There is more to us than brief, solitary lives...plurality of consciousness is only an *appearance.*"

"...subject and object, life and death, you and I, God and man are one....The plain fact is that we are *already* one with the universe; we have never really been apart from it. And only the presence of the 'self' prevents us from seeing...who we *really* are."

Every atom belonging to me as good, belongs to you
—Walt Whitman

You are living in a vast, evidently infinite universe, which scientists are just beginning to investigate in depth.

There is much that is not as yet known about beginnings and endings; cosmic interconnections; the universal field of intelligent omnificence; quantum effects on the brain; and many other areas of mystical proportions.

What the universe, apparently, is telling us is to not take for granted what we *think* we know about the nature of reality. The universe is whispering in our ear, "Remain open to further revelations."

"The cosmic religious experience
is the strongest and noblest
driving force behind scientific
research."

—Albert Einstein

✧

Community

Website, more resources, links
www.livingnonduality.org/science-of-the-sages

Blog
www.livingnonduality.org/self-a-blog.htm

Facebook
www.facebook.com/RobertWolfe.LivingNonduality

Made in the USA
San Bernardino, CA
09 March 2015